LECITHIN

Technological, Biological, and Therapeutic Aspects

ADVANCES IN BEHAVIORAL BIOLOGY

Recent Volumes in this Series

LECITHIN

Technological, Biological, and Therapeutic Aspects

Edited by
ISRAEL HANIN

Loyola University Stritch School of Medicine
Chicago, Illinois

and

G. BRIAN ANSELL

Late of the Medical School
University of Birmingham
Birmingham, United Kingdom

SPRINGER SCIENCE+BUSINESS MEDIA, LLC

Library of Congress Cataloging in Publication Data

International Colloquium on Lecithin (4th: 1986: Chicago, Ill.)
  Lecithin: technological, biological, and therapeutic aspects.
  (Advances in behavioral biology; v. 33)

  "Proceedings of the Fourth International Colloquium on Lecithin, held September 15–17,
1986, in Chicago, Illinois"—T.p. verso.
  Includes bibliographies and index.
  1. Lecithin—Physiological effect—Congresses. 2. Lecithin—Therapeutic use—
Congresses. 3. Lecithin—Analysis—Congresses. I. Hanin, Israel. II. Ansell, G. B. (Gordon
Brian) III. Title. IV. Series. [DNLM: 1. Phosphatidylcholines—congresses. W3 IN7353 4th
1986L / QU 93 I59 1986L]
QP752.L4I58 1986   615'.323'322   87-32811
ISBN 978-1-4757-1935-2

ISBN 978-1-4757-1935-2   ISBN 978-1-4757-1933-8 (eBook)
DOI 10.1007/978-1-4757-1933-8

Proceedings of the Fourth International Colloquium on Lecithin,
held September 15–17, 1986, in Chicago, Illinois

© Springer Science+Business Media New York 1987
Originally published by Plenum Press, New York in 1987
Softcover reprint of the hardcover 1st edition 1987

GORDON BRIAN ANSELL

1927 - 1986

DEDICATION

Brian Ansell was born in Birmingham, England on January 24,
1927. He graduated from King Edward's High School in the Edgbaston
section of Birmingham. He was at Trinity College, Cambridge from
1944 to 1947 where he did the natural sciences tripos with Part I in
Chemistry, Physiology, Zoology and Biochemistry and Part II in
Biochemistry. The B.A. degree was awarded in 1947 and the M.A.
degree in 1951. His education was interrupted from 1947 to 1949 by
national service in the Royal Air Force. From 1949 to 1958 he was in
Cardiff at the Neuropsychiatric Research Center in Whitchurch
Hospital, where he studied for the Ph.D. under Dr. Derek Richter. In
1952 he was awarded the Ph.D. degree by the University of Wales. he
then stayed on as a Senior Biochemist through 1958. Beginning in
1959 he returned to Birmingham as a lecturer in neurochemistry in the
Department of Experimental Psychiatry of the Medical School. He was
promoted to Senior Lecturer in 1964 and to Reader in 1967. At
present this department is the Department of Pharmacology (Pre-
Clinical). While at the University of Birmingham he was awarded the
D.Sc. degree in 1964 in recognition of his very substantial body of
experimental studies.

Brian was extremely active in two societies, The International
Society for Neurochemistry and Biochemical Society. With the
International Society for Neurochemistry he was a member of the

council from 1969 to 1973, then was Treasurer from 1973 to 1977, and was Chairman from 1977 to 1979. After that he was the registered agent because it is incorporated in the United Kingdom. With the Biochemical Society he also served as member of committee, member of the publications committee, chairman of the neurochemical group, archivist, and publications secretary. The latter position which was held for six years included responsibility for the Biochemical Journal and all other publications of the Biochemical Society.

For the Journal of Neurochemistry, Brian was a member of the editorial board from 1965 to 1972 and then Deputy-Chief Editor until 1976. He was involved in the organization of the international neurochemistry meeting held in Oxford in 1965 and was local organizer of a satellite meeting of the International Society for Neurochemistry on Phospholipid Metabolism in the Nervous System in 1981. Together with J.N. (Tim) Hawthorne, he produced books on phospholipids in 1964, 1973, and 1982, that were published by Elsevier in Amsterdam. These books have been very important for the development of research in the area of phospholipids and their metabolism.

The research of Brian Ansell began with the studies of proteases and amino acids. He then began studies of water soluble ethanolamine compounds including phosphoethanolamine and glycerophosphoethanolamine. These led him into the study of phospholipid turnover. He did important work in the early 1960s on plasmalogens and glycerol ethers, which we now know as the ether-linked glycerophospholipids. It was as a result of reading these papers that I applied for a post-doctoral fellowship to study with Brian. He was particularly concerned with function of the brain and the relationship of choline phospholipids with the cholinergic system. He was the first to study the choline phosphotransferase for the synthesis of choline glycerophospholipids from CDPcholine and is well known for his numerous studies dealing with choline phospholipids. During his last visit to the United States in September of 1986, he discussed his current work on the phosphodiesterase acting on glycerophosphocholine.

Sheila Spanner joined Brain Ansell's laboratory in 1959 as a technician. She completed her academic degrees through the Ph.D. degree by thesis and examination with research work done in the laboratory. I will always treasure the year that I was privileged to spend in the laboratory with Brian and Sheila form 1964-65. Numerous other students and post-doctoral fellows such as Tadeusz Chojnacki from Warsaw also have had the opportunity to learn with Brian and Sheila.

Brian had two primary hobbies. The first of these was cinema and films. He certainly would have enjoyed a period of retirement with all of the video tapes of the classic films that are available today. While I was in Birmingham I became interested in collecting coins. We were soon in the habit of checking our change each day for new dates to add to our collections.

Brian passed away in November of 1986, just six weeks after the meeting in Chicago. He is survived by his widow Edwina and two children, Caroline and Christopher. The sudden loss of Brian reminds us again to treasure our friends and loved ones while they are with us and we can enjoy them.

Lloyd A. Horrocks

PREFACE

Recently, there has been tremendous scientific interest in the
role of phospholipids and particularly phosphatidylcholine (lecithin)
in a variety of biological processes. These include the involvement
of phosphatidylcholine in biological membranes, as a component of
plasma lipoprotein, as a transporter of choline in the body, and also
as a "reserve", and possibly only source of unesterified choline.
Moreover, numerous clinical studies have recently been conducted, to
evaluate the possible uses of externally supplied lecithin in the
treatment of some intractible neuropsychiatric disorders (e.g.
tardive dyskinesia, Alzheimer's disease, etc.) and other conditions
(e.g. hypercholesteremia).

Results to date are encouraging, yet equivocal. This is due, in
part, to the fact that the field of phospholipid methodology is
highly complex. There is much confusion in the literature, and many
ambiguities still remain in the interpretation of experimental
findings. This is particularly so for phenomena involving
phospholipid function in the central nervous system.

This book incorporates the proceedings of the Fourth
International Colloquium on Lecithin, which took place in Chicago,
Illinois, USA, on September 15-17, 1986. The purpose of this
colloquium was to review, in a comprehensive manner, basic principles
as well as current information about the technology, biochemistry,
physiology and therapeutic potential of lecithin. Over 108
individuals from all over the world participated in the sessions.
The meeting was subdivided into oral presentations and panel
discussions (see Table of Contents). It was sponsored and financed
by the Lucas Meyer GmbH (Hamburg), for which the editors are most
grateful.

The scientifically stimulating and exciting atmosphere generated
during the colloquium was, subsequently, marred by the death of
Dr. G. Brian Ansell, less than two months after this meeting, as the
result of a heart attack on November 20th, 1986. Brian had had a
prior heart attack about a year earlier, but he felt well enough to
continue to be involved in the planning phases of this colloquium,
and he flew in to the United States from England, in order to
participate in the colloquium. During the meeting he participated
actively in all the proceedings and was, as always, his usual witty,
alert and effusive self. Most of his friends and colleagues who saw
Brian for the last time at this meeting will therefore remember him
as such. I, personally, feel a keen sense of loss. Dr. Lloyd
Horrocks, in his tribute which follows, expresses so effectively the
feelings that I am sure many of Brian's friends and scientific
colleagues felt when they found out about his sudden death.

Dr. G. Brian Ansell has been a lifelong contributor to our understanding of many aspects of phospholipid metabolism and function. This was the last major undertaking which he was involved in before his death. He worked so diligently in assuring that all aspects of the planning of this conference, and bringing it to fruition, were achieved. Furthermore, the success of this conference was extremely important to him. It therefore is befitting that this book should be dedicated to his memory. He will be both remembered, and missed.

Israel Hanin

Chicago, IL   USA

CONTENTS

I. TECHNOLOGY

II BIOLOGY

III. THERAPEUTIC CONSIDERATIONS

IV. PANEL DISCUSSIONS

OVERVIEW ON PHOSPHOLIPIDS: CHEMISTRY, NOMENCLATURE

AND ANALYTICAL METHODOLOGY

G.B. Ansell

Department of Pharmacology, The Medical School,
University of Birmingham, Birmingham B15 2TJ. U.K.

After decades in which phospholipids were considered as
inconvenient, impure mixtures of intractable materials, claiming the
attention of only a few devotees (Klenk, Levene) the era commencing in
about 1950 has been one of intense fundamental research on their
nature, metabolism and function in animals, micro-organisms and plants.

New methods of synthesis pioneered by Baer, Malkin and van Deenen
were paralleled by new methods of analysis, first an ingenious
technique using partial hydrolysis introduced by Dawson, and then a
flood of chromatographic methods involving separations on columns of
alumina, silica and cellulose. There is now a vast armamentarium
including thin layer chromatography, gas-liquid chromatography and high
performance liquid chromatography. It is likely that few if any novel
phospholipids remain to be discovered and attention is now focussed on
the role of phospholipids in living systems.

The recommendations for the nomenclature of phospholipids proposed
in 1976 by the IUPAC-IUB Commission on Biochemical Nomenclature are
rather complicated but at least there is some order now.
Glycerophospholipids are derivatives of sn-glycerol 3-phosphate (e.g.
1-palmitoyl-2-stearoyl-sn-glycero-3-phosphocholine) and sphingophospho-
lipids are derivatives of 4-E-sphingenine (sphingosine). Some of the
nomenclature is idiosyncratic, as is that for plasmalogens and a
cursory glance at the literature will disclose some "consumer
resistance".

Though the size of phospholipid molecules are of the same order
there are infinite, subtle modifications of the long chain fatty acids
and aldehydes which must be related to their structural as well as
metabolic roles. Our knowledge of the special functions of
phospholipids containing alkyl, alkenyl groups or carbon-phosphorus
bonds is still scanty as is a role for sphingosine. It is a curious
irony that a phospholipid whose structure was established only in the
early nineteen sixties, phosphatidylinositol 4,5,-bisphosphate, now has
a firmly established role as an integral component of a "second
messenger system". Furthermore, the hydrophilic component of the
molecule serves as one messenger and the hydrophobic component another.
There is, in short, a whole new era in phospholipid research before us.

MODERN TECHNIQUES FOR THE FRACTIONATION AND PURIFICATION OF PHOSPHOLIPIDS

FROM BIOLOGICAL MATERIALS

Lloyd A. Horrocks, Laura L. Dugan, Cheryl J. Flynn,
Gianfrancesco Goracci, Serena Porcellati, and Young Yeo

The Ohio State University
Department of Physiological Chemistry
1645 Neil Avenue, 214 Hamilton Hall
Columbus, Ohio 43210

INTRODUCTION

Glycerophospholipids have important roles in biological processes. In addition to forming the hydrophobic backbone of cell membranes, glycerophospholipids also participate actively in signal transduction. A number of receptors respond to agonist-binding by activation of a phospholipase C that hydrolyzes PtdIns $4,5$-P_2 to $InsP_3$ and diacylglycerols (Berridge and Irvine, 1984; Berridge et al., 1985; Berridge, 1986; Nishizuka, 1984a, b; Hirasawa and Nishizuka, 1985; Abdel-Latif et al., 1985; Akhtar and Abdel-Latif, 1986). The $InsP_3$ may increase cytosolic Ca^{2+} concentrations, thus activating a protein kinase and other reactions. The diacylglycerol stimulates protein kinase C.

Choline glycerophospholipids may also be involved in signal transduction (Snyder, 1985). Several cell types form platelet activating factor (Lee et al., 1986; Benveniste, 1985; Oda et al., 1985; Venuti, 1985). This begins with a receptor-associated hydrolysis of 1-alkyl-2-arachidonoyl-sn-glycero-3-phosphocholine (see Fig. 1) to produce 1-alkyl-2-lyso-GroPCho and free arachidonic acid, the precursor of numerous eicosanoids (Albert and Snyder, 1983; Touqui et al., 1985; Sanchez-Crespo et al., 1985). The lyso PakCho is then acetylated at the 2-position to produce platelet activating factor (Fig. 1). This takes place in most inflammatory cells. Platelet activating factor not only activates platelets, but also stimulates neutrophils to release leukotriene B_4 (Poitevin et al., 1985). Smooth muscle has platelet activating factor receptors (Doyle et al., 1986; Yousufzai and Abdel-Latif, 1985). Platelet activating factor is involved in anaphylactic shock, histamine release, acute inflammation, graft rejection, and gastrointestinal ulcerations (Bourgain et al., 1985; Orlov et al., 1985; Hayashi et al., 1985; Saeki et al., 1985; Feuerstein et al., 1985; Hwang et al., 1986; Braquet et al., 1985). In the perfused rat liver, platelet activating factor mediates glycogenolysis and vasoconstriction (Buxton et al., 1986a, b). The reactions and enzymes of ether-linked glycerophospholipids have been reviewed (Snyder, 1985; Hanahan, 1986).

3

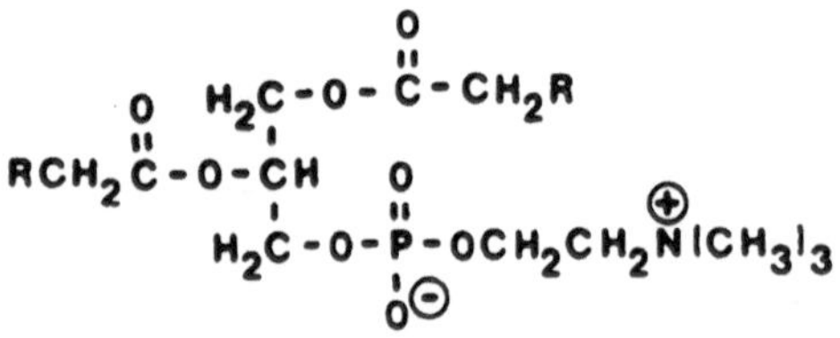

1,2–diacyl–sn–Gro–3–PCho
(phosphatidylcholine, PtdCho)

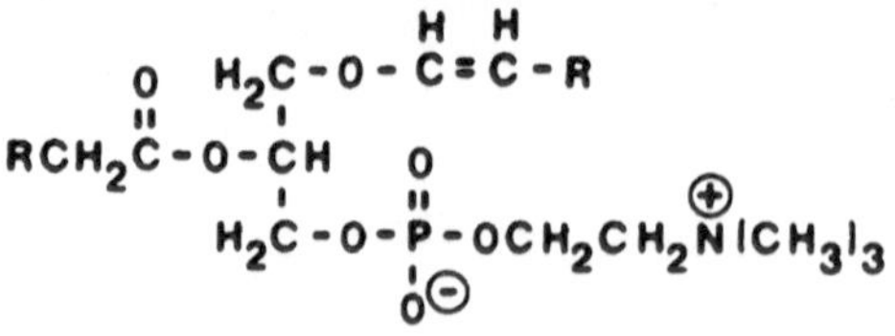

1–alk–1'–enyl–2–acyl–sn–Gro–3–PCho
(plasmenylcholine, PlsCho)

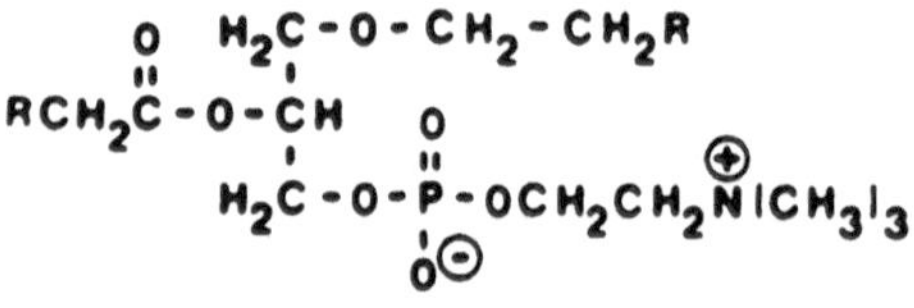

1–alkyl–2–acyl–sn–Gro–3–PCho
(alkylacyl–GroPCho, PakCho)

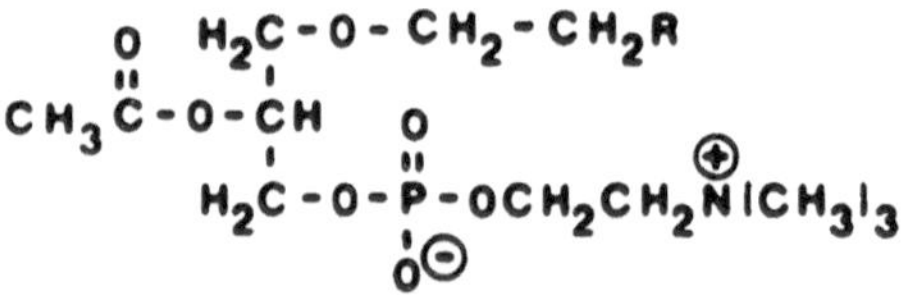

1–alkyl–2–acetyl–sn–Gro–3–PCho
(platelet activating factor)

Fig. 1. Nomenclature of choline glycerophospholipids. Several systems of nomenclature are used for the choline glycerophospholipids. The 1,2-diacyl-sn-Gro-3-PCho (utilizing the abbreviations suggested by the IUB-IUPAC) has the common name phosphatidylcholine. The latter should be used only for the diacyl type according to the recommendations of the IUB-IUPAC, but the term phosphatidylcholine and the abbreviation PC are often used for the mixture of all three types of choline glycerophospholipids. PC is ambiguous because it can mean phosphocholine (PCho) or phosphatidylcholine (PtdCho).

The common name for 1-alk-1'-enyl-2-acyl-sn-Gro-3-PCho is choline plasmalogen. The recommended name, plasmenylcholine, is not yet used very often but can be useful. We suggest the abbreviation PlsCho for plasmenylcholine. No common name exists for 1-alkyl-2-acyl-sn-Gro-3-PCho.

The nomenclature commission suggested plasmanylcholine but in our opinion this is too easily confused with plasmenylcholine. Without coining a new term, the shortest distinctive name is alkylacyl-GroPCho. We suggest the abbreviation PakCho to indicate a choline glycerophospholipid with an alkyl group. This is consistent with the name "phosphalkanylcholine" proposed by P. Karlson of Marburg in 1974 (letter to members of the working group on the nomenclature of lipids).

Choline plasmalogens (Fig. 1) are another source for the receptor-mediated release of arachidonic acid and subsequent formation of eicosanoids (Horrocks et al., 1986a, b). Platelet aggregation by thrombin includes the metabolism of arachidonic acid from choline and ethanolamine plasmalogens (Fig. 2). Human platelets were prelabeled with [^{3}H]arachidonic acid, then stimulated with a low dose of thrombin. During the first 20 sec, the radioactivity in the PlsCho increased 25%, then decreased. Between 20 sec and 3 min, the radioactivity in PlsEtn doubled.

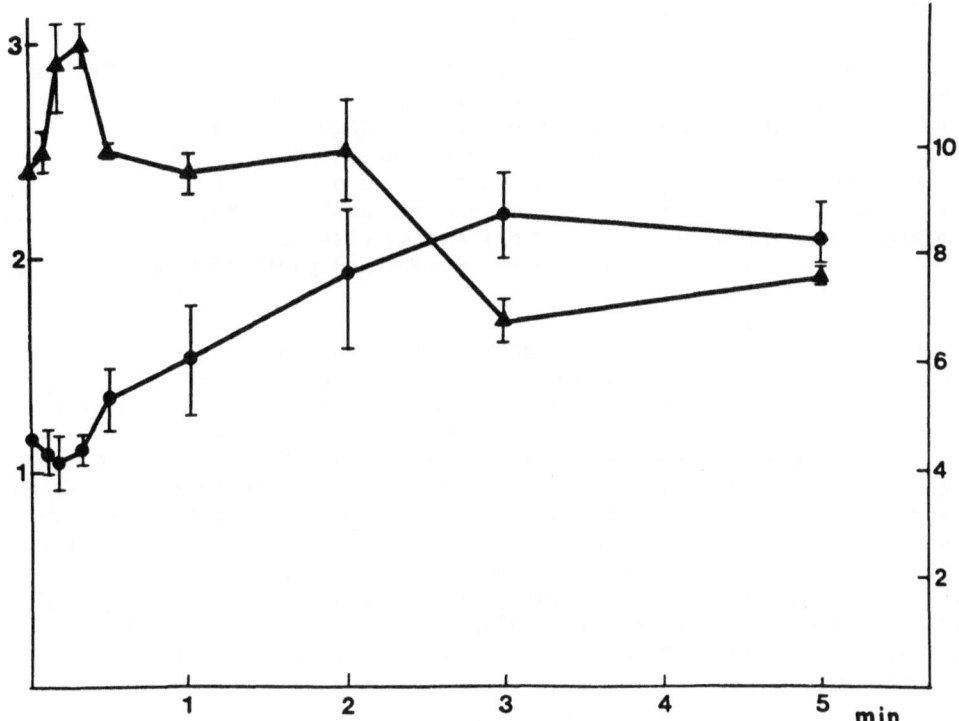

Fig. 2. Effect of stimulation with thrombin on the radioactivity of
plasmalogens in platelets prelabeled with arachidonic acid. Human
platelets were prelabeled with [³H]arachidonic acid (Porcellati et al.,
1986) and washed (Kinlough-Rathbone et al., 1983) except that 1 μM PGI₂
was also added to the medium. The platelet concentration was adjusted to
$2 \cdot 10^5$ cells·ml-¹ in Tyrode-albumin buffer.

Platelets, 0.5 ml, were stimulated with thrombin, 0.5 U·ml-¹, and the
distribution of the radioactivity was determined at the indicated times.
The filled triangles and left-hand scale are for PlsCho and the filled
circles and right-hand scale are for PlsEtn, percent radioactivity.

The changes in levels of radioactivity indicate hydrolysis and
reacylation of the plasmalogens. The ether-linked glycerophospholipids
incorporate arachidonic acid at the 2-position by transarachidonoylation
from the 2-position of PtdCho (Malone et al., 1985; Sugiura and Waku,
1985; McKean and Silver, 1985; Sugiura et al., 1985; Robinson and Snyder,
1985; Colard et al., 1986). The transfer is energy independent. The role
of plasmalogens as important sources of free arachidonic acid is often
overlooked because the active metabolism is masked by the
transarachidonoylation. The ether-linked lysoglycerophospholipids may
also be involved in secretory responses such as glucose-induced insulin
secretion (Metz, 1986). The very early hydrolysis of the choline
plasmalogens may be linked with the entry of external Ca^{2+}.

A massive influx of extracellular Ca^{2+} into cells is found immediately
after trauma of the spinal cord (Stokes et al., 1983; Stokes et al., 1984;
Stokes et al., 1985; Young and Flamm, 1982; Young and Koreh, 1986). The
largest change in the glycerophospholipids at that time is the loss of
10-18% of the PlsEtn which comprises more than one-third of the membrane
phospholipids (Demediuk et al., 1985). Polyphosphoinositides were not
examined.

Muscle tissues contain relatively large amounts of both choline and ethanolamine plasmalogens (Horrocks, 1972; Horrocks and Sharma, 1982; Gross, 1985; Arthur et al., 1985; Kostetskii and Sergeyuk, 1985). Values for bovine muscles are 19% PlsCho and 21% PlsEtn in the total phospholipids. Electrical stimulation of the tissue caused a loss of 18-20% of both plasmalogens with no significant change in any other glycerophospholipid. Polyphosphoinositides were not examined in this study. The stimulation of the beef muscle may activate phospholipases associated with receptors. A phospholipase A_2 with specificity for plasmalogens has been detected in heart muscle and purified from platelets (Wolf and Gross, 1985; Loeb and Gross, 1986).

Glycerophospholipids with alkyl groups have antitumor activities (Andreesen et al., 1978, 1982; Berdel et al., 1981). Synthetic derivatives of 2-lyso PakCho cause the accumulation of alkyl-containing glycerophospholipids in tumor cells because they are deficient in the alkyl-cleavage enzyme. Normal cells metabolize the compounds and are not affected. The N-acyl derivatives of EtnGpl accumulate in infarcted canine myocardium and cerebrum (Natarajan, 1981; Natarajan et al., 1980) and in degenerating chick embryo tissues in fertilized hens' eggs (Kara et al., 1986). A tumor-selective cytolytic phospholipid preparation from the latter source was fractionated. The active component was the N-palmitoyl derivative of PakEtn (Kara et al., 1986). It is non-toxic for normal cells but inhibits DNA synthesis by malignant cells at nanomolar concentrations.

The differences in metabolism of glycerophospholipids with acyl, alkenyl, or alkyl groups at the 1-position indicate a requirement for analytical methods to separate all classes. Marked metabolic differences are also found for different molecular species within specific classes. For example, the turnover of PlsCho, PakCho, and PtdCho with 18:0 side-chains at the 1-position is much slower than the turnover of the compounds differing only in having an 18:1 side-chain (Horrocks et al., 1986a, b). In addition, the polyphosphoinositides, free fatty acids, and diacylglycerols should also be extracted and quantitated. The accepted methods and their limitations are reviewed in much greater detail in a book from the Neuromethods series (Boulton et al., 1987).

Most studies of glycerophospholipids and related products associated with signal transduction have focused on polyphosphoinositides and have used acidic extraction solvents. The addition of HCl to solvents seems to be necessary for removal of the polyphosphoinositides from interactions with the proteins (Hauser and Eichberg, 1973). Unfortunately, these strongly acidic solvents also produce large amounts of artifactual diacylglycerols and free fatty acids (Table 1). In addition, all of the plasmalogens are hydrolyzed to produce long-chain aldehydes and lysoglycerophospholipids.

Table 1. Effects of hydrochloric acid on lipid extraction from rat brain.

	Neutral	Acidified
	nmol lipid per g wet weight, mean ± S.E.M. (n=6)	
PtdIns 4-P	137 ± 5	228 ± 13
PtdIns 4,5-P_2	177 ± 7	572 ± 22
Free fatty acid	161 ± 12	1100 ± 137
Diacylglycerols	312 ± 23	714 ± 94

Rats were killed by microwave irradiation. The brains were extracted with 2:1 chloroform/methanol (Folch et al., 1957) or with 0.5 M HCl also included (T.S. Reddy, L.A. Horrocks, and N.G. Bazan, unpublished).

Mixtures of chloroform and methanol are most often used for the extraction of lipids (Folch et al., 1957; Bligh and Dyer, 1959). These extracts include proteolipid proteins that are difficult to remove. The non-lipid contaminants are often removed by washing procedures that also cause losses of lysoglycerophospholipids, prostaglandins, and lipoxygenase products. The use of chloroform in the laboratory is undesirable because it may induce hepatomas in users.

A mixture of hexane and 2-propanol has been introduced for lipid extraction (Hara and Radin, 1978). The extracts contain low amounts of non-lipid contaminants and are thus suitable for HPLC separations without washing. Nearly all glycerophospholipids are extracted quantitatively from brain tissue (Saunders and Horrocks, 1984) with the exception of polyphosphoinositides. From 25 to 40% of the polyphosphoinositides are not extracted. Preliminary results indicate that the inclusion of additional water in the hexane/2-propanol enables the quantitative extraction of the polyphosphoinositides from brain tissue and cell cultures without using acid (Dugan et al., 1986b). This modification of the previous method (Hara and Radin, 1978) may enable the simultaneous extraction of all of the lipid compounds that are important in signal transduction without generating any artifacts.

For many purposes it is desirable to separate the lipid groups in the extracts. This can be done with a small column of Unisil silicic acid from Clarkson Chemical Co., Williamsport, PA (Saunders and Horrocks, 1984). This method is similar to previous methods that have used chloroform, acetone, and methanol for the elution of lipid groups, but also includes a new second fraction eluted with methyl formate. The chloroform fraction includes neutral lipids and free fatty acids, the methyl formate fraction includes the prostaglandins, most of the lipoxygenase products, and some glycolipids, the acetone fraction has the remainder of the glycolipids and lipoxygenase products, and the methanol fraction includes the phospholipids. It is also possible to proceed directly to HPLC with a separation of the phospholipid classes (Dugan, et al., 1986a). For the resolution of neutral lipids and free fatty acids with this HPLC method, it is necessary to collect them and reinject them for another HPLC separation.

Most investigators are now separating phospholipids by TLC or HPTLC methodology. Excellent 1-dimensional separations of phospholipids can be obtained (Gustavsson, 1986). She has developed reproducible methods for photodensitometry that are independent of the degree of saturation of the

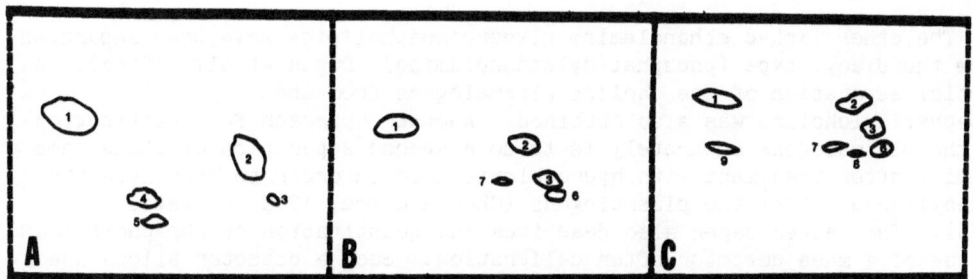

Fig. 3. Thin-layer chromatograms of Krebs cell phospholipids following no treatment (A), phospholipase A_1 treatment (B), and phospholipase A_1 and acid treatments (C) (El Tamer et al., 1984). The separation was on silica gel H with 4% magnesium trisilicate. A basic solvent with chloroform, methanol, and ammonia was used first (up), and the second solvent was acidic with chloroform, methanol, and dilute acetic acid (right to left). The solvent sequence was reversed for C.

phospholipids. As mentioned above, it is usually desirable to separate the plasmalogens from the corresponding glycerophospholipids. This can be done easily by 2-dimensional TLC with an exposure to the fumes of hydrochloric acid between dimensions (Horrocks, 1968; Horrocks and Sun, 1972; Saunders and Horrocks, 1984). Excellent separations of the glycerophospholipid classes, but without separation of the plasmalogens, can be obtained with a two-dimensional procedure (Rouser et al., 1966).

All three types of glycerophospholipid classes can be quantitated with a complex TLC method (Fig. 3) (El Tamer et al., 1984). This method utilizes 2-dimensional TLC separations with exposure to hydrochloric acid fumes between dimensions. In addition, one portion of the lipid extract is treated with a phospholipase A_1 before chromatography in order to remove the diacyl components. Suitable enzymes with phospholipase A_1 activity include Rhizopus arrhizus preparations and purified pancreatic lipase. After the removal of all of the diacyl types, the remaining material that is acid-stable is the alkylacyl type, whereas the acid-labile material is derived from the alkenylacyl type. The amount of diacyl compound is obtained by subtracting the value for acid-stable materials before and after the phospholipase A_1 hydrolysis.

We have developed a number of separation techniques using high performance liquid chromatography for the separation and quantitation of tissue lipids. The separations are based on the use of the solvent system containing n-hexane/2-propanol/water (Hax, 1977). Some of these methods have been reported recently (Dugan et al., 1986a). All of the major classes of glycerophospholipids are separated. Recently, we have found that a longer separation using the same solvents enables the separation of platelet activating factor, phosphatidylinositol 4-phosphate and phosphatidylinositol 4,5-bisphosphate (Fig. 4). Previously, for the quantitation of phospholipid classes we have used separations by thin layer chromatography with scraping of the spots and ashing for the determination of phosphorus. Separation of the phospholipids by high performance liquid chromatography gives better recoveries of the phospholipids and improved reproducibility of the phosphorus determination. Thus smaller changes in the phospholipid composition can be detected with statistical significance. A high performance liquid chromatography method for the separation of all of the methylated phospholipids has also been devised (Chen and Kou, 1982). It is also possible to program the solvents so that greater separation of the less polar lipids is achieved. In this way, free fatty acids and each of the phospholipid classes in extracts from brain were separated with a single high performance liquid chromatography procedure.

The ether linked ethanolamine glycerophospholipids have been separated from the diacyl type (phosphatidylethanolamine) (Dugan et al., 1986a). A partial separation of the choline plasmalogens from the phosphatidylcholine was also obtained. Another approach for quantitation of the plasmalogens separately is to do a second separation of these same samples after treatment with hydrochloric acid in order to hydrolyze the alkenyl groups from the plasmalogens (Chen and Kou, 1982; Christie, 1985). The latter paper also describes the quantitation of phospholipids by use of a mass detector after calibration. Such a detector allows use of solvents such as chloroform that cannot be used with ultraviolet detectors. Christie's separation included all of the principal neutral lipid and phospholipid classes. The separation of the molecular species of the intact glycerophospholipids is very difficult. After removal of the 2-acyl groups, some of the 1-radyl molecular species of choline and ethanolamine glycerophospholipids can be separated by reversed-phase HPLC (Creer and Gross, 1985).

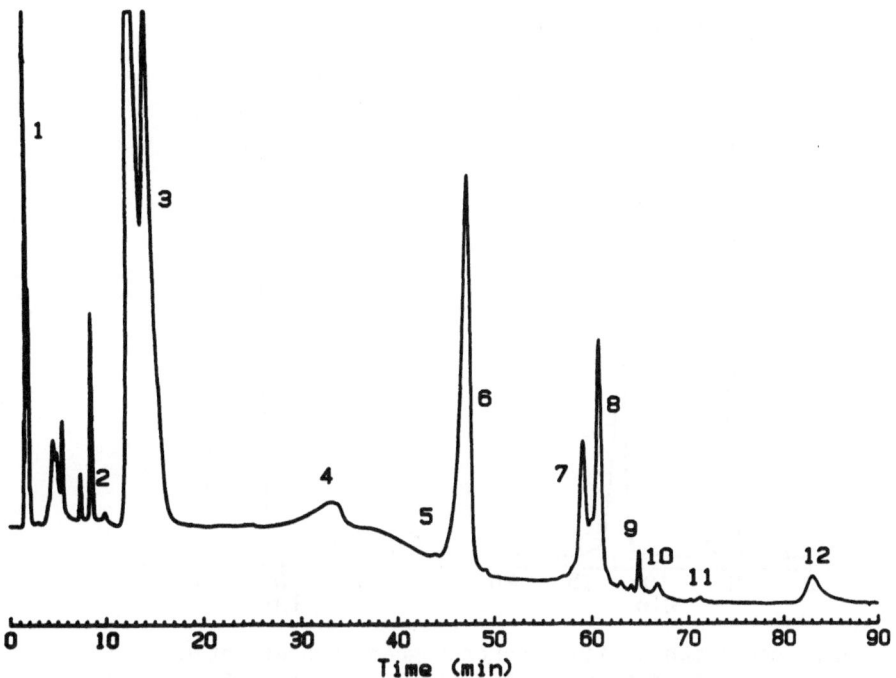

Fig. 4. High-performance liquid chromatogram of the separation of rat
brain phospholipids. The separation was done as described previously
(Dugan et al., 1986a), but with about 20 additional minutes of elution.
The peaks are 1, neutral lipids; 2, cardiolipin; 3, EtnGpl with partial
separation of the plasmalogens in the first peak; 4, phosphatidylinositol;
5, lysoEtnGpl; 6, phosphatidylserine; 7 and 8, ChoGpl with partial
separation of the plasmalogens in the first peak; 9, sphingomyelin; 10,
phosphatidylinositol 4-phosphate; 11, lysoChoGpl; and 12,
phosphatidylinositol 4,5-bisphosphate.

The intact diglyceride portions of the phospholipids are required for
many investigations. We have developed the high performance liquid
chromatography separation of three types of these diglycerides from
glycerophospholipids, namely the diacyl, alkylacyl, and alkenylacyl types
(Nakagawa and Horrocks, 1983). The 3-acetyl derivatives were made in
order to prevent migration of the 2-acyl group to the 3-position. From
1-50 μmoles of these acetyl derivatives were separated at one time with
detection at 205 nm. This separation was used for the diglycerides from
the choline and ethanolamine glycerophospholipids from mouse brain
(Fig.5). The amounts of the diglycerides were quantitated by collection
of the eluted fractions, preparation of the fatty acid methyl esters, and
assay of the fatty acids using an internal standard. The EtnGpl contained
41% alkenylacyl, 4% alkylacyl, and 55% diacyl types, whereas the ChoGpl
contained only 1.0% alkenylacyl and 1.9% alkylacyl types, with the bulk in
the diacyl type.

The direct quantitation of these three types of diglycerides requires
the use of a derivative with the same absorbance for each type and
molecular species. Blank et al. have prepared benzoates of these three
types and have separated them by thin-layer chromatography. After
scraping and extracting the silica gel, they quantitated by absorption at

HPLC Separation of Alkenylacyl, Alkylacyl, and Diacyl GPE and GPC

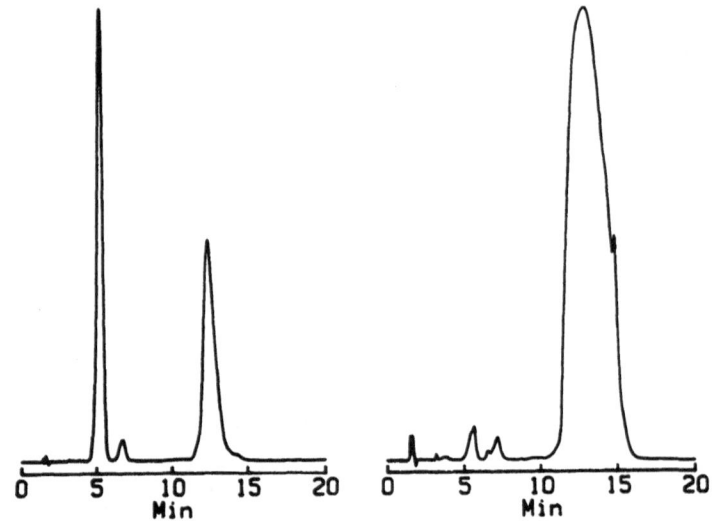

Fig. 5. High performance liquid chromatogram of acetylated diradylglycerols derived from the EtnGpl (left) and ChoGpl (right) of mouse brain. The method was described previously (Nakagawa and Horrocks, 1983). A μPorasil silicic acid column (Waters) with a particle size of 10 μm was used with a solvent system of cyclopentane/hexane/methyl t-butyl ether/glacial acetic acid in a ratio of 70:27:3:0.03 at a flow rate of 2 ml·min⁻¹ at 33° C. Detection was at 205 nm.

230 nm (Blank et al., 1984). They also showed that the molecular species could be separated by reverse phase high performance liquid chromatography. For this, 30 nmoles of each type were needed. We have used dinitrobenzoate derivatives of the diglycerides because 5 nmoles are sufficient for the high performance liquid chromatography separation and quantitation of the three types. This derivative can also be used for the separation of 5-30 nmoles of various molecular species (Kito et al., 1985). In order to detect the liberation of diglycerides during physiological and pathological processes, even more sensitive methods are required. We are now investigating the use of naphthoyl derivatives with detection by fluorescence. Fig. 6 shows the separation of such derivatives by high performance liquid chromatography with sample sizes of 100 picomoles of the diacyl and alkenylacyl types.

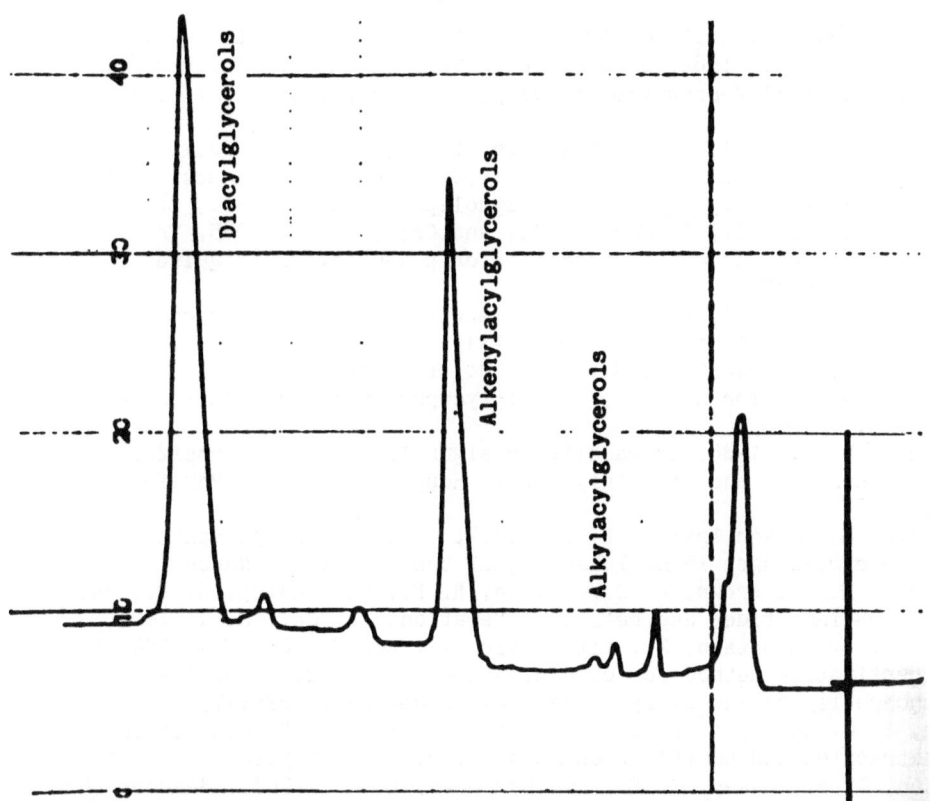

Fig. 6. High performance liquid chromatogram of the naphthoyl derivatives of diradylglycerols derived from the ChoGpl of bovine heart. The latter were hydrolyzed with phospholipase C from Bacillus cereus. The diradylglycerols were reacted with an excess of naphthoyl chloride in dry pyridine for 2 hours at 60° C. After solvent washings, an aliquot containing about 200 pmol of the naphthoyl derivatives was injected. A 5 cm guard column with Viosil Si60 and a 25 cm analytical column with Viosfer Si 10 μm (Violet, Rome, Italy) were used with a gradient solvent system. The solvents were cyclohexane/hexane/2-propanol in a ratio of 60:40:0.1 with an additional 0.5% methyl t-butyl ether in solvent A. The proportion of A was increased from 3% to 15% during the run. The excitation was at 285 nm and the detection at 365 nm with a fluorescence detector.

REFERENCES

Abdel-Latif, A. A., Smith, J. P., and Akhtar, R. A., 1985, Polyphosphoinositides and muscarinic cholinergic and alpha-adrenergic receptors in the iris smooth muscle., in: "Inositol and Phosphoinositides: Metabolism and regulation.," Bleasdale, J. E., Eichberg, J., and Hauser, G., eds., Humana Press, Clifton, NJ.

Akhtar, R. A. and Abdel-Latif, A. A., 1986, Surgical sympathetic denervation increases alpha 1-adrenoceptor-mediated accumulation of myo-inositol trisphosphate and muscle contraction in rabbit iris dilator smooth muscle., J. Neurochem., 46:96.

Albert, D. H. and Snyder, F., 1983, Biosynthesis of 1-alkyl-2-acetyl-sn-glycerol-3-phosphocholine (platelet-activating factor) from 1-alkyl-2-acyl-sn-glycero-3-phosphocholine by rat alveolar macrophages. Phospholipase A2 and acetyltransferase activities during phagocytosis and ionophore stimulation, J. Biol. Chem., 258:97.

Andreesen, R., Modolell, M., Oepke, G. H., Common, H., Lohr, G. W., and
    Munder, P. G., 1982, Studies on various parameters influencing
    leukemic cell destruction by alkyl-lysophospholipids, Anticancer Res.,
    2:95.
Andreesen, R., Modolell, M., Weltzien, H. U., Eibl, H., Common, H. H.,
    Lohr, G. W., and Munder, P. G., 1978, Selective destruction of human
    leukemic cells by alkyl-lysophospholipids, Cancer Res., 38:3894.
Arthur, G., Mock, T., Zaborniak, C., and Choy, P. C., 1985, The
    distribution and acyl composition of plasmalogens in guinea pig heart,
    Lipids, 20:693.
Benveniste, J., 1985, PAF acether (Platelet activating factor), Adv.
    Prostaglandin Thromboxane Res., 13:11.
Berdel, W. E., Bausert, W. R., Fink, U., Rastetter, J., and Munder, P. G.,
    1981, Anti-tumor action of alkyl-lysophospholipids, Anticancer Res.,
    1:345.
Berridge, M. J., 1986, Intracellular signalling through inositol
    trisphosphate and diacylglycerol., Hoppe-Seyler's Z. Physiol. Chem.,
    367:447.
Berridge, M. J. and Irvine, R. F., 1984, Inositol trisphosphate, a novel
    second messenger in cellular signal transduction., Nature, 312:315.
Berridge, M. J., Brown, K. D., Irvine, R. F., and Heslop, J. P., 1985,
    phosphoinositides and cell proliferation., J. Cell Sci., 1985(S3):187.
Blank, M. L., Robinson, M., Fitzgerald, V., and Snyder, F., 1984, Novel
    quantitative method for determination of molecular species of
    phospholipids and diglycerides, J. Chromatogr., 298:473.
Bligh, E. G. and Dyer, W. J., 1959, A rapid method of total lipid
    extraction and purification, Can. J. Biochem. Physiol., 37:911.
Boulton, A. A., Baker, G. B., and Horrocks, L. A., 1987, "Neuromethods,"
    vol. 7, Humana Press, Clifton, NJ.
Bourgain, R. H., Touqui, L., Maes, L., Braquet, P., Andries, R., Braquet,
    M., 1985, The effect of 1-0-alkyl-2-acetyl-sn-glycero-3-phosphocholine
    (PAF-Acether) on the arterial wall, Prostaglandins, 30:185.
Braquet, P., Borgeat, P., Etienne, A., and Braquet, M., 1985, Stimulus
    secretion coupling and leukotriene formation in the triggering of
    immediate hypersensitivity reactions, A. I. P. Immunol., 136:186.
Buxton, D. B., Fisher, R. A., Hanahan, D. J., and Olson, M. S., 1986a,
    Platelet activating factor mediated vasoconstriction and
    glycogenolysis in the perfused rat liver, J. Biol. Chem., 261:644.
Buxton, D. B., Hanahan, D. J., and Olson, M. S., 1986b, Specific
    antagonists of platelet activating factor mediated vasoconstriction
    and glycogenolysis in the perfused rat liver, Biochem. Pharmacol.,
    35:893.
Chen, S. F. and Chan, P. H., 1985, One step separation of free fatty acids
    and phospholipids in brain tissue extracts by high performance liquid
    chromatography, J. Chromatogr., 344:297.
Chen, S. S. and Kou, A. Y., 1982, High-performance liquid chromatography
    of methylated phospholipids., J. Chromatogr., 232:237.
Christie, W. W., 1985, Rapid separation and quantification of lipid
    classes by HPLC and mass (light scattering) detection, J. Lipid Res.,
    26:507.
Colard, O., Breton, M., and Bereziat, G., 1986, Arachidonate mobilization
    in diacyl, alkylacyl and alkenylacyl phospholipids on stimulation of
    rat platelets by thrombin and the Ca2+ ionophore A23187, Biochem. J.,
    233:691.
Creer, M. H. and Gross, R. W., 1985, Reversed phase high performance
    liquid chromatographic separation of molecular species of alkyl ether,
    vinyl ether, and monoacyl lysophospholipids, J. Chromatogr., 338:61.
Demediuk, P., Saunders, R. D., Anderson, D. K., Means, E. D., and
    Horrocks, L. A., 1985, Membrane lipid changes in laminectomized and
    traumatized cat spinal cord, Proc. Nat. Acad. Sci. U.S.A., 82:7071.

Doyle, V. M., Creba, J. A., and Ruegg, U. T., 1986, Platelet activating factor mobilizes intracellular calcium in vascular smooth muscle cells, FEBS Lett., 197:13.

Dugan, L. L., Demediuk, P., Pendley II, C. E., and Horrocks, L. A., 1986a, Separation of phospholipids by HPLC: All major classes, including ethanolamine and choline plasmalogens, and most minor classes, including lysophosphatidylethanolamine., J. Chromatog., 378:317.

Dugan, L., Bazan, N. G., and Horrocks, L. A., 1986b, Extraction/HPLC separation of polyphosphoinositides by neutral solvents, Trans. Amer. Soc. Neurochem., 17:138.

El Tamer, A., Record, M., Fouvel, J., Chap, H., and Douste-Blazy, L., 1984, Studies on ether phospholipids .1. A new method of determination using phospholipase-A1 from guinea-pig pancreas - Application to Krebs-II Ascites-cells, Biochim. Biophys. Acta, 793:213.

Feuerstein, G., Lux Jr., W. E., Ezra, D., Hayes, E. C., Snyder, F., and Faden, A. I., 1985, Thyrotropin-releasing hormone blocks the hypotensive effects of platelet-activating factor in the unanesthetized guinea pig, J. Cardiovasc. Pharmacol., 7:335.

Folch, J., Lees, M., and Sloane Stanley, G. H., 1957, A simple method for the isolation and purification of total lipides from animal tissues, J. Biol. Chem., 226:497.

Gross, R. W., 1985, Identification of plasmalogen as the major phospholipid constituent of cardiac sarcoplasmic reticulum, Biochemistry, 24:1662.

Gustavsson, L., 1986, Densitometric quantification of individual phospholipids. Improvement and evaluation of a method using molybdenum blue reagent for detection, J. Chromatogr., 375:255.

Hanahan, D. J., 1986, Platelet activating factor: a biologically active phosphoglyceride, Annu. Rev. Biochem., 55:483.

Hara, A. and Radin, N. S., 1978, Lipid extraction of tissues with a low-toxicity solvent, Anal. Biochem., 90:420.

Hauser, G. and Eichberg, J., 1973, Improved conditions for the preservation and extraction of polyphosphoinositides, Biochim. Biophys. Acta, 326:201.

Hax, W. M. A., 1977, High-performance liquid chromatographic separation and photometric detection of phospholipids, J. Chromatogr., 142:735.

Hayashi, H., Kudo, I., Inoue, K., Onozaki, K., Tsushima, S., Nomura, H., and Nojima, S., 1985, Activation of guinea pig peritoneal macrophages by platelet activating factor (PAF) and its agonists, J. Biochem., 97:1737.

Hirasawa, K. and Nishizuka, Y., 1985, Phosphatidylinositol turnover in receptors: mechanism and signal transduction, Ann. Rev. Pharmacol. Toxicol., 25:147.

Horrocks, L. A., 1968, The alk-1-enyl group content of mammalian myelin phosphoglycerides by quantitative two-dimensional thin-layer chromatography, J. Lipid Res., 9:469.

Horrocks, L. A., 1972, Content, composition, and metabolism of mammalian and avian lipids that contain ether groups, in: "Ether Lipids; Chemistry and Biology," Snyder, F., ed., Academic Press, New York.

Horrocks, L. A. and Sharma, M., 1982, "Phospholipids, New Comprehensive Biochemistry," Hawthorne, J. N. and Ansell, G. B., eds., vol. 4, Elsevier Biomedical Press, Amsterdam.

Horrocks, L. A. and Sun, G. Y., 1972, Ethanolamine plasmalogens, in: "Research Methods in Neurochemistry," Rodnight, R. and Marks, N., eds., Plenum Press, New York.

Horrocks, L. A., Harder, H. W., Mozzi, R., Goracci, G., Francescangeli, E., Porcellati, S., and Nenci, G. G., 1986a, "Enzymes of Lipid Metabolism," Freysz, L. and Gatt, S., eds., vol. 2, Plenum Press, New York.

Horrocks, L. A., Yeo, Y. Y., Harder, H. W., Mozzi, R., and Goracci, G., 1986b, Choline plasmalogens, glycerophospholipid methylation, and receptor-mediated activation of adenylate cyclase, Adv. Cyclic Nucleotide Protein Phosphorylation Res., 20:263.

Hwang, S. B., Lam, M. H., Li, C. L., and Shen, T. Y., 1986, Release of platelet activating factor and its involvement in the 1st phase of carrageenin induced rat foot edema, Eur. J. Pharmacol., 120:33.

Kara, J., Brorvicka, M., Liebl, V., Smolikova, J., and Ubik, K., 1986, A novel nontoxic alkyl-phospholipid with selective antitumor activity, plasmanyl-(N-acyl)-ethanolamine (PNAE), isolated from degenerating chick embryonal tissues and from an anticancer biopreparation cACPL, Neoplasma, 33:198.

Kinlough-Rathbone, J. D., Packham, M. A., and Mustard, J. F., 1983, Measurement of platelet function, in: "Methods in Haematology," Harker, L. A. and Zimmerman, T. S., eds., Churchill-Livingstone, Edinburgh.

Kito, M., Takamura, H., Narita, H., and Urade, R., 1985, A sensitive method for quantitative analysis of phospholipid molecular species by HPLC, J. Biochem., 98:327.

Kostetskii, E. Y. and Sergeyuk, N. N., 1985, Phospholipids and their plasmalogen form in the muscle tissue of marine invertebrates, J. Evol. Biochem., 21:133.

Lee, T. C., Malone, B., and Snyder, F., 1986, A new de novo pathway for the formation of 1-alkyl-2-acetyl-sn- glycerols, precursors of platelet activating factor. Biochemical characterization of 1-alkyl-2-lyso-sn-glycero-3-P:acetyl CoA acetyltransferase in rat spleen., J. Biol. Chem., 261:5373.

Loeb, L. A. and Gross, R. W., 1986, Identification and purification of sheep platelet phospholipase A2 isoforms, J. Biol. Chem., 261:10467.

Malone, B., Lee, T. C., and Snyder, F., 1985, Inactivation of platelet activating factor by rabbit platelets. Lyso-platelet activating factor as a key intermediate with phosphatidylcholine as the source of arachidonic acid in its conversion to a tetraenoic acylated product, J. Biol. Chem., 260:1531.

McKean, M. L. and Silver, M. J., 1985, Phospholipid biosynthesis in human platelets. The acylation of lyso platelet activating factor, Biochem. J., 225:723.

Metz, S. A., 1986, Ether-linked lysophospholipids initiate insulin secretion. Lysophospholipids may mediate effects of phospholipase A2 activation on hormone release, Diabetes, 35:808.

Nakagawa, Y. and Horrocks, L. A., 1983, Separation of alkenylacyl, alkylacyl, and diacyl analogues and their molecular species by high performance liquid chromatography, J. Lipid Res., 24:1268.

Natarajan, V., 1981, On the biosynthesis and metabolism of N-acylethanolamine phospholipids in infarcted dog heart, Biochim. Biophys. Acta, 774:445.

Natarajan, V., Epps, D. E., Schmid, P. C., and Schmid, H. H. O., 1980, Accumulation of N-acylethanolamine glycerophospholipids in infarcted myocardium, Biochim. Biophys. Acta, 618:420.

Nishizuka, Y., 1984a, The role of protein kinase C in cell surface signal transduction and tumor promotion., Nature, 308:693.

Nishizuka, Y., 1984b, Turnover of inositol phospholipids and signal transduction., Science, 225:1365.

Oda, M., Satouchi, K., Yasunaga, K., and Saito, K., 1985, Production of platelet activating factor by washed rabbit platelets, J. Lipid Res., 26:1294.

Orlov, S. M., Kulikov, V. I., Polner, A. A., and Bergelson, L. D., 1985, Release of histamine from human leukocytes induced by 1-O-alkyl-2-O-acetyl-sn-glycero-3-phosphocholine (platelet activating factor) and its structural analogs, Biochemistry-USSR, 50:575.

Poitevin, B., Mencia-Huerta, J. M., Roubin, R., and Benveniste, J., 1985, Role of PAF acether (Platelet Activating Factor) in neutrophil activation, in: "Pulmonary Circulation and Acute Lung Injury," Said, S. I., ed., Future Publishing Co.

Porcellati, S., Goracci, G., Costantini, V., Pistolesi, R., Nenci, G. G., and Horrocks, L. A., 1986, October, Plasmalogens are metabolized in platelets during thrombin aggregation, Paper presented at Second International Conference, Platelet-Activating Factor and Structurally-Related Alkyl Ether Lipids, Gatlinburg, TN.

Robinson, M. and Snyder, F., 1985, Metabolism of platelet activating factor by rat alveolar macrophages. Lyso PAF as an obligatory intermediate in the formation of alkylarachidonoyl glycerophosphocholine species, Biochim. Biophys. Acta, 837:52.

Rouser, G., Siakotos, A. N., and Fleischer, S., 1966, Quantitative analysis of phospholipids by thin-layer chromatography and phosphorus analysis of spots, Lipids, 1:85.

Saeki, S., Musugi, F., Ogihara, T., Otsuka, A., Kumahara, Y., Watanabe, K., Tamura, K., Akashi, A., and Kumagai, A., 1985, Effects of 1-0-alkyl-2-acetyl-sn-glycero-3-phosphocholine (Platelet activating factor) on cardiac function in perfused guinea pig heart, Life Sci., 37:325.

Sanchez-Crespo, M., Alonso, F., Garcia-Gil, M., Gomez-Cambrone, J., and Neito, M. L., 1985, Synthesis of platelet activating factor from human polymorphonuclear leukocytes. Regulation and pharmacological approaches, Int. J. Tissue Res., 7:345.

Saunders, R. D. and Horrocks, L. A., 1984, Simultaneous extraction and preparation for high performance liquid chromatography of prostaglandins and phospholipids, Anal. Biochem., 143:71.

Snyder, F., 1985, Chemical and biochemical aspects of platelet activating factor: a novel class of acetylated ether-linked choline phospholipids, Med. Res. Rev., 5:107.

Stokes, B. T., Fox, P., and Hollinden, G., 1983, Extracellular calcium activity in the injured spinal cord, Exp. Neurol., 80:561.

Stokes, B. T., Fox, P., and Hollinden, G., 1985, Extracellular metabolites: Their measurement and role in the acute phase of spinal cord injury, in: "Trauma of the Central Nervous System," Dacey, J. R. G. E. A., ed., Raven Press, New York.

Stokes, B. T., Hollinden, G., and Fox, P., 1984, Improvement in injury induced hypocalcia by high-dose naloxone intervention, Brain Res., 290:187.

Sugiura, T. and Waku, K., 1985, CoA independent transfer of arachidonic acid from 1,2-diacyl-sn-glycero-3-phosphocholine to 1-0-alkyl-sn-glycero-3-phosphocholine (lyso platelet-activating factor) by macrophage microsomes, Biochem. Biophys. Res. Commun., 127:384.

Sugiura, T., Masuzawa, Y., and Waku, K., 1985, Transacylation of 1-0-alkyl-sn-glycero-3-phosphocholine (lyso platelet-activating factor) and 1-0-alkenyl-sn-glycero-3-phosphoethanolamine with docosahexaenoic acid (22:6ω3), Biochem. Biophys. Res. Commun., 133:574.

Touqui, L., Jacquemin, C., Dumarey, C., and Vargaftig, B. B., 1985, 1-0-Alkyl-2-acyl-sn-glycero-3-phosphorylcholine is the precursor of platelet activating factor in stimulated rabbit platelets. Evidence for an alkylacetyl glycerophosphorylcholine cycle, Biochim. Biophys. Acta, 833:111.

Venuti, M. C., 1985, Platelet activating factor. Multifaceted biochemical and physiological mediator, Annu. Rep. Med. Chem., 20:193.

Wolf, R. A. and Gross, R. W., 1985, Identification of neutral active phospholipase C which hydrolyzes choline glycerophospholipids and plasmalogen selective phospholipase A2 in canine myocardium, J. Biol. Chem., 260:7295.

Young, W. and Flamm, E. S., 1982, Effect of high-dose corticosteroid therapy on blood flow, evoked potentials, and extracellular calcium in experimental spinal injury, J. Neurosurg., 57:667.

Young, W. and Koreh, I., 1986, Potassium and calcium changes in injured spinal cords, Brain Res., 365:42.

Yousufzai, S. Y. and Abdel-Latif, A. A., 1985, Effects of platelet activating factor on the release of arachidonic acid and prostaglandins by rabbit iris smooth muscle, Fed. Proc., 44:488.

MICROORGANISMS AS SOURCES OF

PHOSPHOLIPIDS

Colin Ratledge

Department of Biochemistry
University of Hull
Hull HU6 7RX, U.K.

INTRODUCTION

Almost the entire range of plant and animal phospholipids can be found within microorganisms. The range has been comprehensively reviewed by Ambron and Pieringer (1973) and more recently by Thiele (1979) in an all-encompassing monograph on lipids. More selective reviews include those by Pieringer (1983) covering principally the biosynthesis of bacterial phospholipids, Weete (1980) dealing with fungal lipids and Mangnall and Getz (1973) reviewing eukaryotic (yeasts, fungi and algae) lipids. Other reviews which may be consulted include those by Brennan and Lösel (1978), covering specialized areas of fungal lipids; and by Verma and Khuller (1983) covering lipids of Actinomycetes. As a most useful introductory text to the subject of microbial (and plant) lipids, the monograph by Harwood and Russell (1984) can be recommended.

Although micro-organisms may be considered to be a potential source of many lipids, including phospholipids, only one commercial process presently exists for this purpose and that is for the production of a fungal triacylglycerol oil rich in γ-linolenic acid. (This is reviewed in slightly more detail in the section devoted to Fungal Phospholipids.) As far as this author is aware, no specific process exists for the extraction of phospholipids from any microbial source even when that source may be available as a waste material from a biotechnology process devoted to the production of some other material. There have, however, been some interests expressed in this area though these do not appear to have reached commercial take-up and there is at least one process which produces phospholipids in a mixture of other lipids. Again, this is discussed in more detail in the Section on Yeast Phospholipids.

In this chapter, I propose to review the overall potential of micro-organisms as sources of phospholipids from a practical standpoint. That is to assess if there are any sources of microbial biomass which could be used as starting materials for the extraction of these lipids, and if so, what would be their nature. There are two possible approaches which might be taken, and indeed both will be considered here.

Firstly, as phospholipids are intrinsically associated with membranous structures, we could expect to find higher than average amounts of phospholipids to occur in microorganisms which possess an abundance of membranes.

The principal organisms which then fall into this category would be (i) photosynthetic organisms, that is mainly algae and cyanobacteria (blue-green algae), where the photosynthetic apparatus would be the main source of membrane phospholipids; (ii) methylotrophic bacteria, that is bacteria grown on methane or methanol as sole carbon source, where the membranes are associated with the process of C_1-assimilation; and (iii) yeasts grown on hydrocarbons which also engenders formation of additional organelles. The second strategy for examining potential sources of phospholipids is to look at those microorganisms which are grown in bulk for some other purpose and whose biomass may then be available at little or no cost. Obvious examples here include yeast (Saccharomyces cerevisiae) recovered from brewing processes, and moulds such as Aspergillus niger or Penicillium spp. recovered from citric acid or antibiotic productions.

To some extent the two strategies are found to be complementary in that algae, methylotrophic bacteria and hydrocarbon-grown yeast are, or until recently have been, available as commercial sources of human or animal feed material. It is therefore probably most sensible to review separately each of the four major categories of microorganism - algae, bacteria, mould and yeast - and to make use of the co-incidence that phospholipids may be in some abundance in at least some organisms already being cultivated on a large scale. [Details concerning the large scale cultivation of microorganisms have been extensively reviewed in recent years as part of the explosion of interest in biotechnology. The treatises edited by Rehm and Reed (1985) and by Moo-Young (1985) may be recommended as being particularly comprehensive should the erudite reader require further information of this topic.]

LIPID FORMATION IN MICROORGANISMS IN GENERAL

The amount of lipid which microorganisms may accumulate can be up to, and even slightly in excess of, 80% of the dry weight of the microbial cell. Accumulation of lipid is, however, not universal amongst micro-organisms and most in fact accumulate no more than 5 to 10% of their weight as lipid, no matter how they are grown. Organisms which do accumulate lipid above a somewhat arbitrary level of 20% of the biomass have been termed oleaginous (Whitworth and Ratledge, 1974) in keeping with the same term used to describe high oil-yielding plant seeds.

The lipid which is stored by an oleaginous microorganism is almost invariably a neutral lipid. This can be a wax-ester or polyester (such as poly-β-hydroxybutyrate in bacteria) but is more usually a triacyl-glycerol as in yeasts, moulds and algae. Occasionally, hydrocarbons may accumulate in a few algae. The amount of phospholipid which is found in such oleaginous cells is rarely more than is found in non-oleaginous microorganisms. Thus, as is shown later, the amount of lipid within a cell does not increase the amount of phospholipid which then normally fall within the range of 1 to 6% of the biomass (see Tables 6 and 9).

The principal factor which governs the accumulation of lipid in oleaginous microorganisms is that the growth medium for the organism should be low in an essential nutrient other than carbon. Usually this is the amount of nitrogen being added. The organism when grown on such a medium quickly exhausts the supply of nitrogen but continues to assimilate the excess carbon supply from the medium. This carbon, usually it is carbo-hydrate, is then converted into lipid. The lipid is therefore regarded as a storage compound produced when carbon is in abundance and capable of being utilized if the organism subsequently becomes starved. Few bacteria may be regarded as oleaginous; this property is then found in a small number of yeasts, a greater number of moulds and also in some algae (Ratledge, 1982; 1984; 1986).

The process of lipid accumulation depends on the control of the central pathways of metabolism ensuring a continual supply of acetyl-CoA and NADPH by which fatty acid biosynthesis may proceed without hindrance. Additional factors then come into play which govern the degree of unsaturation of the fatty acids, their chain-length and, to some extent, whether lipids other than triacylglycerols, the principal lipid storage form, are also synthesized. These factors will then be expected to influence not only the amount of phospholipid being synthesized but also the various types which are present as well as the fatty acyl substituents. Such factors include the growth rate of the organism, its growth temperature, supply of oxygen, presence (or absence) of salts, choice of carbon substrate and its concentration, choice of nitrogen source (NH_4^+ or an organic source), growth pH etc. These effects often bring about changes in different directions and it is sometimes difficult to interpret the results with any certainty. For example, if no precautions are taken to control the rate at which the organism is growing - which can be done using a chemostat - then the effect of growth temperature on, say, the fatty acyl composition of the organism becomes a complex event. A change in the growth temperature will invariably alter the growth rate of the organism and will also change the dissolved O_2 concentration. A direct interpretation of the effect of the initial variables (temperature) has then become impossible. A discussion of some of these factors which can influence lipid synthesis has been presented elsewhere (Boulton and Ratledge, 1985) and is also reviewed in considerable detail for bacterial lipids by Lechevalier (1982).

MICROALGAL PHOSPHOLIPIDS

Algae are a diverse collection of organisms ranging from the macro-algae of the red and brown seaweeds to the microalgae which includes both prokaryotic and eukaryotic forms. The former group is composed of the cyanobacteria, previously known as the blue-green algae. (It should though be noted that there are photosynthetic bacteria other than the cyanobacteria: these are the Anoxyphotobacteria, which are not regarded as microalgae. Their lipid compositions are those of other Gram-positive bacteria.) Both groups of microalgae have representatives which are cultivated on a large scale as sources of human and animal food as well as potential sources of biologically active compounds (Metting and Pyne, 1986). Algae, which may be either freshwater or marine, can be rich sources of lipids (Wood, 1974; Pohl, 1982). The lipids though tend to be of a wide variety of types, many of which may be unusual (Mangold, 1986).

The extent of lipid formation can be controlled by the growth conditions ensuring a low level of NH_4^+ or NO_3^- in the environment: up to 75% of Chlorella pyrenoidosa can be lipid (Milner, 1951) if grown under such conditions. Several reviews have appeared summarizing the potential of algae as source of lipids: Shifrin and Chisholm (1980); Shifrin (1984); Aaronson et al. (1980); Materassi et al. (1980); Ciferri and Tiboni (1985). The main algae which are considered for commercial development, mainly as food or feed material but which may be suitable as sources of lipid, include Chlorella (see Milner, 1951; and also Prokop and Fekri, 1984), Oscillatoria (Yanagimoto and Saitoh, 1982) Spirulina (Ciferri and Tiboni, 1985) Dunaliella - for growth in hyper-saline environments such as the Dead Sea (Dubinsky et al., 1978) and Neochloris (Tornabene et al., 1983b). Of these Spirulina spp. are currently produced in four locations: Miyako Island in Japan, Republic of China, Thailand and Mexico. It is also recovered as a mixed culture of algae from Lake Chad, Africa.

The lipids extracted from various algae are almost invariably complex mixtures containing besides neutral lipids (which includes sterols and sterol esters), glycolipids and phospholipids, numerous pigments associated with the thylakoid membranes of photosynthetic apparatus: chlorophylls, carotenes and quinones. The range and properties of the phospholipids are given in Tables 1 and 2. Prokaryotic algae, as exampled by Spirulina maxima in Table 2, tend to produce phosphatidylglycerol as the main, if not sole, phospholipid; this is usually not more than 8% of the total lipids (see also Zebke et al., 1978). Although the lipid content of Spirulina maxima, as a commercial sample from Lake Texcoco, Mexico, was calculated as 11% by Hudson and Karis (1974), there is evidently much variation in this value as Yanagimoto and Saitoh (1982) have quoted values as little as 1.4 and 2.6% for the lipid contents of commerical Spirulina samples. In eukaryotic algae, as exampled by Dunaliella bardawil and Neochloris oleoabundans in Table 2, the types of phospholipids are as found in other eukaryotic organisms.

The fatty acyl groups of the phospholipids, as well as the other lipid fractions, have been frequently studied (see above refs. and Wood, 1974) and these include polyunsaturated fatty acids in both the prokaryotic and eukaryotic algae. Such fatty acids include 16:2, 16:3, 18:2, α-18:2, γ-18:3, 20:4 and 20:5 (Oren et al., 1985).

BACTERIAL PHOSPHOLIPIDS

The lipids of bacteria are predominantly comprised of phospholipids and these are localized almost entirely within the various membrane structures of the cell. For details concerning this topic, the reviews of Goldfine (1972), Finnerty (1978) and Pieringer (1983) can be recommended.

The cytoplasmic membranes of Gram-positive bacteria (Micrococcaceae,

Table 1. Lipid composition of some algae used in large scale cultivation

	% Lipid of biomass	% Lipid composition			
		Neutral	Phospho-*	Polar†	Ref.
Spirulina maxima (ex Lake Texcoco, Mexico)	11	10	30	70	1
Spirulina platensis**	16.6	5	← 95% →		2
Neochloris oleoabundans**	54	91	6	3	3
Dunaliella bardawil**	30	50	6	44	4

  * see Table 2
  † mainly mono- and digalactosyldiacylglycerols
** grown to promote lipid accumulation

Refs. 1, Hudson and Karis (1974); 2, Tornabene et al. (1983a)
       3, Tornabene et al. (1983b); 4, Fried et al. (1982)

Table 2. Phospholipids of selected algae (see Table 1)

Component	Rel. percent in		
	D. bardawil	N. oleo-abundans	S. maxima
Phosphatidylcholine	34	10	-
Phosphatidylethanolamine	22	35	-
Phosphatidylglycerol	44	21	84
Phosphatidylinositol	-	5	16*
Phosphatidylserine	-	15	-
Diphosphatidylglycerol	-	16	-

* identity queried

(data from references quoted in Table 1)

Streptococcaceae, Peptococcaceae, Bacillaceae and Lactobacillaceae) consist in most species of phosphatidylglycerol and its derivatives, including O-aminoacyl esters. Phosphatidylethanolamine (PE) and diphosphatidylglycerol (DPG) - cardiolipin - are less common though in Gram-negative bacteria PE is by far the most abundant and common phospholipid. Phosphatidylcholine (PC) is comparatively rare in all bacteria though it can be found in the photosynthetic membranes of purple and green bacteria (Rhodospirilliceae and Chromatiaceae).

In the archaebacteria, which includes the halophiles (e.g. Halobacterium) and thermophiles (e.g. Thermus, Thermoplasma), the major phospholipids of the former group are the diether analogues of phosphatidylglycerophosphate (PGP) and PG. Related di- and tetra-ether lipids (see Harwood and Russell, 1984) are found in the thermophiles.

The only group of bacteria to contain more than a few percent of their biomass as lipid, are the actinomycetes where the lipid contents may be up to 40% of the dry weight of some organisms such as the tubercle bacillus (Mycobacterium tuberculosis). However these lipids are complex and are frequently either toxic or shown to have immunological activities (Ratledge, 1976; Goren and Brennan, 1979). In general, phospholipids may form about 2 to 4% of the dry weight of mycobacteria and related genera (see Verma and Khuller, 1983). PE, PG and DPG are present in most actinomycetes (the main genera being Mycobacterium, Nocardia, Corynebacterium, Streptomyces and Rhodococcus) with PS and lyso-PE often being minor components. Within the entire order of the actinomycetes, PC has only been reported in a few strains of Streptomyces and also in Nocardia coeliaca (Verma and Khuller, 1983). The principal phospholipids of the mycobacteria and some other genera are, however, the phosphatidylinositol mannosides (PIMs) (Verma and Khuller, 1983; Goren and Brennan, 1979; Minnikin and O'Donnell, 1984) where up to six mannoses, of which five may be as a single oligosaccharide,may be attached to the inositol ring. Related PIMs but with fewer mannose residues are found also in the propionibacteria.

Despite the wide assay of phospholipid types within the bacterial kingdom, potential sources of these materials are limited by the very availability of the bacteria themselves. Furthermore, as many bacteria may possess toxic or immunoreactive lipids, such as the lipopolysaccharides of Gram-negative bacteria, opportunities for producing bacterial lipids are limited to those organisms which have undergone some toxicological trials prior to their incorporation into food. Of course, many bacteria are used in conjunction with the production of certain foods: yoghurt, cheese, fermented meats which involve principally lactobacilli and propionibacteria. The bacteria, however in these examples, are not separated from the food itself but become an intrinsic part of it. Thus we are left only with those bacteria which are grown as sources of single cell protein (SCP).

The subject of biotechnology and the production of SCP has often been reviewed, with several complete monographs being presented on the subject. The reviews of Norris (1981), and Vasey and Powell (1984) and the books of Rose (1976) and Goldberg (1985) will probably be sufficient to give an adequate background. Most SCP processes tend to use yeasts or fungi as the production organism; the only commercial bacterial SCP processes are those involving methane or methanol as carbon substrate. Of these, methanol has been the preferred choice. Whilst yeasts and bacteria will both grow on methanol (only bacteria will grow on methane), the higher growth rate of the bacteria coupled with a higher yield (g cells per g substrate) has made bacteria the organism of choice for SCP production from methanol(Anthony, 1982; - chp. 12 therein). Three of the best known industrial processes are those developed by I.C.I. plc (U.K.); Hoechst GmbH (W. Germany) and Mitsubishi Gas Chemical Co. (Japan). All use, or have used, strains of Methylomonas or Methylophilus. The compositions of the bacterial biomasses are given in Table 3. It will be seen that the amount of lipid in the various biomasses is between 8 and 10%. The lipid is predominantly composed of phospholipid (up to 90-92% - see Patt and Hanson, 1978) as the bacterial cells possess extensive internal membrane structures. Thus between 7 and 8% of the bacterial mass can be phospholipid; these microorganisms therefore provide the highest amounts of phospholipid in the entire microbial kingdom.

Methylotrophic bacteria are divided into two categories according to their biochemical characteristics. This division also coincides with differences in the organization of the internal membranes: Type I methylotrophs (Methylococcus, Methylomonas and Methylobacter) have bundles of disc-shaped vesicles which extend across the cell and appear to be

Table 3. Composition of bacteria grown on methanol as commercial source of SCP

Component	Proximate analysis of samples (% dry matter)		
	1 Methylomonas clara	2 Methylomonas sp. BNK-84	3 Methylophilus methylotrophus
Protein	72	70.8	80*
Nucleic acids	12	14.8	4?
Lipids	9	9.6	8
Ash	7	8.1	8
	100%	103.3%	100%

* crude protein
1. from Faust (1979) for material produced by Hoechst AG, West Germany.
2. from Urakami et al. (1981) for material produced by Mitsubishi, Japan.
3. from Norris (1981) for material produced by I.C.I. plc.

Table 4. Phospholipid composition of methylotrophic
bacteria

A. Type I organisms (used for SCP production)

	PE	PG	DPG	PC
Methylococcus capsulatus[a]	74	13	5	8
Methylomonas methanolica[b]	77	17	6	0

[a] Grown on CH_4 - from Makula (1978)

[b] From Andreev (1978) quoted from Wilkinson (1987)

B. Type II organisms (not used for SCP production)

	PDME	PMME	PG	PC	PE	PS	DPG
Methylosinus trichosporium[a]	49	21	13	11	-	-	-
Methylosinus erichosporium[b]	-	-	57	0.6	38	4	0.4
Methylobacterium organophilum[c]	24	-	-	15	57	1	-
Methylobacterium organophilum[d]	24	-	-	38	25	9	-

[a] Grown on CH_4; from Makula (1978)

[b] Grown on CH_4 (9.2% total lipid = 35% phospholipids); from
Weaver et al. (1975)

[c] Grown on CH_4 (9% total lipid = 88% phospholipids); from
Weaver et al. (1975)

[d] Grown on methanol (6% total lipid = 92% phospholipid); from
Patt and Hanson (1978)

extensions of the cytoplasmic membrane. Type II bacteria (Methylocystis,
Methylosinus and Methylobacterium) have paired peripheral membranes.
(Excellent electron monographs showing the two types of membrane have been
published by Smith and Ribbons, 1970; some of which are also shown in the
monograph of Anthony, 1982.) The membranes can result in a surface area
some eight times that of the cytoplasmic membrane; their function, though,
is still not certain. Type I organisms, having the faster growth rate, are
used in preference to Type II organisms for SCP production. The phospholipid composition of both Type I and Type II organisms is given in Table 4.

Like most other bacteria, there is little PC present in any of these
lipids; the predominant phospholipid is usually PE or its methylated
derivitives, PMME or PDME. The fatty acid composition of the total lipid,
which may be taken to be the same as that of the phospholipids, are given
in Table 5. The range of fatty acids in a number of species are give for
both Type I organisms as well as Type II methylotrophs. Again a distinction
can be clearly seen between the two types; Type I organisms contain predom-

Table 5. Fatty acid composition of methylotrophic
bacteria (from Wilkinson, 1978)

A.   Type I organisms (used for SCP production)

	14:0	16:0	16:1	18:1
Methylococcus spp	7-12%	10-45%	31-74%	0-12%
Methylomonas spp	2-14%	10-40%	30-74%	0-20%
Methylophilus spp	1-2%	32-43%	27-51%	6-8%

B.   Type II bacteria

	16:0	16:1	18:0	18:1*
Methylocystis spp	2-4%	1-8%	1-3%	75-96%
Methylosinus spp	1-8%	2-8%	0-3%	75-88%

* cis-vaccenic acid [cis-18:1 (11)]

inantly palmitoleic acid [cis-16:1(9)]. C_{18} fatty acids rarely occur. In
type II methylotrophs, on the other hand, the amounts of C_{16} acids (16:0
plus 16:1) are low and the lipid consists almost entirely of one fatty acid
- cis-vaccenic acid [cis-18:1(11)].

Due to the highly competitive nature of SCP prices, attempts have been
made by some of the manufacturers to extract out of the biomass various
components which might then have a higher value than the bulk material.
Such considerations have led to an examination of the phospholipid material
from Type I organisms for a variety of applications. Whilst some interest
was undoubtedly expressed in such phospholipids, the economics of extraction
and purification were by no means certain. At any event, the production of
phospholipid was not sufficiently attractive to improve the overall economics
of bacterial SCP production as it now appears that both the European processes have ceased production. (That of ICI plc was considerably more
developed than that of the process of Hoechst, but this too has ceased
operation as of mid-1986.)

Thus the prospects of bacteria as potential sources of phospholipids
have now receded with the disappearance of large scale processes for the
production of bacterial biomass in bulk. With about 7 to 8% of the total
bacterial weight being phospholipid, which has represented one of the best
microbial sources of the material, this demise is all the more unfortunate.

MOULD PHOSPHOLIPIDS

The lipids of moulds have been extensively reviewed by Weete (1980);
the review by Brennan and Lösel (1978) covers selected topics which also
includes a description and discussion of fungal phospholipids. Although
a large number of species are known which accumulate lipid at over 20% of
the biomass (see Ratledge, 1982), this material is usually neutral lipid
and the amount of phospholipid does not increase in proportion. Table 6
gives some of the results of detailed analyses carried out by Suzuki and
co-workers on a large number of moulds.

Table 6. Phospholipids of oleaginous moulds (from Suzuki et al., 1981a,b; 1982a,b)

	% Lipid of cell dry wt.	Polar lipids of cell dry wt.	Rel. % phospholipids*					
			PC	PE	PS	PI	DPG	PG
Mortierella isabellina	84	3	42	17	1	11	-	-
Penicillium lilacinum	51	3	21	22	5	2	5	6
Cladosporium herbarum	49	7	50	16	1	-	18	13
Pellicularia practicola	31	6	52	20	5	3	2	3
Fusarium oxysporum	22	7	25	18	16	20	2	5

* Where figures do not add up to 100%, presence of other polar lipids is indicated.

Even with the highest lipid accumulating species, Mortierella isabellina, the amount of phospholipid was still only 3% of the biomass. Although cultural conditions will alter the amount of lipid being produced, as well as the fatty acid profile of both the neutral and polar lipid fractions, there is little evidence from the literature that high amounts of phospholipid can be deliberately produced. Farag et al. (1983) examined the effect of 13 different growth media on lipid formulation in two moulds, Tolyposporium ehrenbergii and Sphacelotheca reiliana, and only in two instances - both with the latter organism - did the phospholipid content exceed 8% of the dry weight. Even here where the values were about 20% of the cells, there was very sparse growth (<0.6 gl^{-1}) and opportunity for error in the determinations would have been considerable. No indication of reproducibility of these results was given.

The analyses of the various phospholipids (see Table 6 and also data provided by Weete, 1980) indicate that PC is often the most abundant type, though rarely does this exceed 50% of the total. In some moulds, PE may be the most abundant phospholipid (see for example DeBell and Jack, 1975) though it is normally second to PC.

Opportunities for extracting phospholipids from filamentous fungi must be mainly limited to those organisms which are grown commercially for the production of antibiotics or organic acids, such as citric acid. Aspergillus niger is widely cultivated as a commercial source of citric acid and although spent mycelia have been used for the extraction of sterols, principally ergosterol, there is no indication from the literature that any consideration has been given to using this material as a source of phospholipid. Suzuki et al. (1981a) recorded the phospholipid content of four strains of Asp. niger from 1.5 to 2.5% of the biomass (see Table 7). This, however, represented in one case about a third of the total lipid. As with the other fungi (see Table 6), the constituent phospholipids showed a predominance of PC, PE and PS.

Moulds which are being grown as sources of Single Cell Protein may

Table 7. Phospholipids of four strains of
Aspergillus (from Suzuki et al.,
1981a)

Total lipid of mycelium: 8-17%

Phospholipids (% dry biomass): 1.5-2.5%

PC	33-54%
PE	10-33%
PS	7-27%
DPG(CL)	2-15%
PG	0-6%
PI	0-13%
LPC	0-9%

also be of some potential as a source of phospholipids. Currently a large
scale process is being operated by ICI plc (with the same fermenter system
as was used for the production of bacteria SCP - see previous section) on
behalf of Rank, Hovis & MacDougall plc, also of the UK (see Norris, 1981).
This process uses a *Fusarium* sp and at present the mycelia mass is only
processed to remove some of the nucleic acids. The lipid (about 10% of
the mould) is not removed though clearly this could be if the phospholipid
could command sufficient value. The phospholipid composition would be
expected to be similar in composition to those analyses presented in
Tables 6 and 7.

Mycelia from citric acid or SCP production would probably be a better
starting point for phospholipid preparation than moulds used for antibiotic
production where residual biologically-active material could be co-extracted
with the lipids and cause problems of acceptability. Another disadvantage
would be that all antibiotic-producing strains are usually inactivated
before being released from the production plant and furthermore, may
contain high amounts of lipid due to vegetable oils frequently being
incorporated into the production medium (Ratledge, 1977). Unlike citric
acid production, there is little financial incentive for antibiotic
manufacturers to consider extracting any by-product as this is likely to
have little intrinsic value in comparison with that of the primary product.

Some indication of the changes which occur in an antibiotic-producing
fungus is given in Table 8. This may be taken perhaps as typical of the
changes which might be expected in other fungi, including ones not producing antibiotics.

With *Paecilomyces persicinus*, the content of phospholipid was highest
during the growth phase of the organism (see Table 8). It would appear
that little net phospholipid biosynthesis occurred as the cells then grew
from 24 h to 48 h. This probably reflects the route of biosynthesis of the
neutral lipids (presumably mainly triacylglycerols) being from the phospholipids themselves. As the cells aged (from 48 h onwards) there was a steady
increase in proportion of PC with PE showing some decline.

Although the fatty acids of the phospholipids as well as other lipids
have frequently been examined (see Suzuki et al., 1981a,b; 1982a,b;
Ratledge, 1982), there have been few reports of any stereospecific analysis
being carried out. In one notable study, DeBell and Jack (1975) showed
that there are distinct differences between the fatty acyl groups on the
sn-1 and sn-2 positions of PC and PE as extracted from the *Phycomyces*
blakesleeanus. The main differences were the virtual exclusion of 16:0

Table 8. Changes in lipid composition of _Paecilomyces persicinus_ during growth and cephalosporin C production (from Papacharilaon and Pisano, 1984)

Culture age (h)	Cell dry wt ($g l^{-1}$)	Cephalo- sporin production (% maxm)	Lipid (% cell wt)	Rel % lipids		Rel % polar lipids				
				Neutral	Polar	PE	PC	DPG	PS	PA
24	5	-	6.1	55	45	31	21	18	16	14
48	10	84	7.5	79	21	45	28	10	9	8
72	8	100	9.0	80	20	38	31	10	8	13
96	7.3	50	6.5	72	28	33	38	9	8	12
120	6.8	40	6.4	71	29	38	40	10	3	9

from the sn-2 position of both phospholipids. To counter this, there was more than twice the amount of 18:2 on the sn-2 position as on sn-1. Interestingly, 18:3 (that is γ-linolenic acid in this instance) was the same proportion (about 33% of the total acids) at both sn- positions with both PC and PE.

As moulds of the order Mucorales are well known to be producers of γ-linolenic acid, 18:3 (6, 9, 12), rather than the commoner α-linolenic acid, 18:3 (9, 12, 15) which is found in all other fungi as well as most plants, there has been some recent interest in using such organisms for the production of this high value material (Sinden, 1987). With commercial production of such an oil now established by J. & E. Sturge, UK (Anon, 1986) there is a prospect that phospholipids with a high content of γ-linolenic acid might be extractable from the residual biomass after the primary extraction of the triacylglycerol oil has taken place. Such phospholipids could then be expected to show the same high proportion of γ-linolenic acid, at 33% of the total fatty acids, as DeBell and Jack (1975) had found with Phycomyces blakesleeanus. Such phospholipids may then have a much higher value than conventional materials.

YEAST PHOSPHOLIPIDS

The lipids of yeasts have been extensively reviewed by Rattray (Rattray et al., 1975; Rattray, 1987) and are also included in reviews on fungal lipids (q.v.). Details concerning the possibilities of using yeasts as sources of Single Cell Oil have been reviewed by the present author (Ratledge, 1982, 1984, 1986). The involvement of lipids in the structure and function of the yeast membranes has been reviewed by Prasad (1984) and Henry (1982).

Although several yeasts are known to be able to accumulate relatively high amounts of lipid, the amount of phsopholipid does not increase in proportion to the total lipid content (see Table 9). Indeed like moulds, it may be better to consider those yeasts with a relatively low lipid content as the better potential source of phospholipid as problems of purification and removal of the neutral lipid might present much less difficulty with a higher phospholipid to total lipid ratio. The values quoted for phospholipid contents in Table 9 are typical for some 30 yeasts which Kaneko et al. (1976) examined when the phospholipid proportion of the biomass ranged from 2.4% to 6.0% and total lipid contents varied from 6.3% to 32.3% (all values being repeated 5 to 10 times on yeasts grown 2 to 4 times).

Table 9. The amount of lipid within a yeast cell does not change the phospholipid content (from Kaneko et al., 1976)

	Total lipid*	PL*	PL/Total lipid
Sacch. cerevisiae	7.2%	3.7%	52%
Sacch. carlsbergensis	7.1%	4.4%	62%
Candida utilis	10.6%	4.0%	38%
Trichosporon cutaneum	13.3%	5.3%	40%
Rhodotorula rubra	18.6%	3.0%	16%
Lipomyces starkeyi	32.3%	5.3%	15%

* Percent of dry biomass

Table 10. Summary of typical phospholipid
compositions of yeast (from
Hunter and Rose, 1971; Weete,
1980; Rattray, 1987)

Phosphatidylcholine (PC)	35-55%
Phosphatidylethanolamine (PE)	25-32%
Phosphatidylinositol (PI)	9-22%
Phosphatidylserine (PS)	4-18%
Diphosphatidylglycerol (DPG) (cardiolipin)	1-4%
Phosphatidylglycerol (PG)	0-3%
Phosphatidic acid (PA)	0-10%

The phospholipid types within various yeasts, whether ascosporogenous
(e.g. Saccharomyces, Hansenula, Lipomyces) or basidiomycetous (e.g. Candida,
Rhodosporidium, Rhodotorula, Trichosporon) are broadly similar (see Table
10). PC is usually the most predominant phospholipid with PE and PI being
the second and third (the exact order between these three types can
occasionally change).

It will be readily appreciated because of the economic importance of
S. cerevisiae in both the brewing and baking industries that the majority
of yeast lipid analyses have been carried out with this species (which
now includes S. carlsbergensis, S. uvarum, S. saké and others). However
even within one single species where there are so many strains and variants,
the variability in phospholipid composition is almost as great as is seen
in the entire remainder of the yeasts (Rattray, 1987).

Numerous examples could be cited of the effects of different growth
conditions and of different nutrients on the lipid and phospholipid composition of S. cerevisiae though mainly for economy of space only a few can
be given here. Hunter and Rose (1972) showed some time ago that changing
the growth temperature at a constant growth rate had little effect on the
total phospholipid content of the cells (approx 4% of the dry weight).
Even changes in the fatty acid composition were not significantly changed
between 15°C and 30°C. However the fatty acid composition of the membrane
lipids can be altered by adding various exogenous fatty acids, particularly
under anaerobic conditions, when S. cerevisiae becomes dependent upon a
supply of oleic acid (or alternative unsaturated fatty acid) and ergosterol
as neither can be synthesized in the absence of oxygen (Alterthum and Rose,
1983; Hossak and Rose, 1976; Thomas et al., 1979; see also Henry, 1982).
Tolerance to alcohol, which of course is often a pre-occupation in studies
with S.cerevisiae, probably depends more on the nature of the phospholipids
though the substituent fatty acids may clearly be of importance.

Variations in the phospholipid composition of S. carlsbergensis (now
S. cerevisiae) have been reported by including thiamine or thiamine plus
pyridoxine in the growth medium (Nishikawa et al., 1977) or by growing the
cells in the complete absence of inositol under anaerobic conditions (Daum
et al., 1977). In the latter case, not unexpectedly the proportion of PI
in the phospholipid fraction was greatly diminished (from about 24% to about
5%); no single other phospholipid however then appeared as a replacement.
A more unexpected result, however, was obtained by Ramsay and Douglas (1979)
who found that growing S. cerevisiae in phosphate-limiting medium in a
chemostat did not in fact diminish the phospholipid content of the cells.
What did diminish was the phosphomannan component of the cell wall, though

29

phosphate-limited cells still had 71% of the total phosphorus content compared to that of control cells.

Although a number of yeasts (but not S. cerevisiae) will grow on hydrocarbons, and will usually increase their lipid contents by so doing (Ratledge, 1980), there is only a slight increase in the amount of phospholipid - though if this is expressed as a percentage increase then one may see as much as a trebling of the phospholipid percentage of the dry weight (see Table 11). This, as in methanol-grown bacteria (q.v.), is attributable to the increased membrane structures which can be seen in such yeasts (Boulton and Ratledge, 1984). The fatty acids of the lipids from alkane-grown yeast correspond closely to chain length of the substrate (Ratledge, 1980) and therefore this will reflect, in turn, changes in the fatty acyl substituents of the various phospholipids.

Yeasts which are grown commercially and are suitable for human or animal consumption are restricted to S. cerevisiae and Candida utilis, the latter being grown in a number of instances on waste carbohydrates as a source of Single Cell Protein (Rose, 1976; Norris, 1981; Goldberg, 1985). Neither appears to be used as a source of phospholipid, presumably because of the relatively low content (see Table 9). Possibilities do though exist for extracting phospholipid out of spent brewer's yeast as is already done in one or two instances for the production of ergosterol. Usually, this yeast is available at little cost.

At the moment, no oleaginous yeast is being grown as a source of Single Cell Oil, though there may be certain prospects in view. Such yeasts could, if grown on a large scale, perhaps generate phospholipids as a desirable by-product. (This point was discussed in the previous section in relation to the current commercial production of γ-linolenic acid in moulds.) However there has been one quasi-commercial process based on yeasts which has generated phospholipids in some quantity though these are unsuitable for human or animal consumption or applications.

This process is based on a process in the USSR and East Germany for the growth of yeasts on hydrocarbons as a source of Single Cell Protein. The yeasts, probably Candida lipolytica, C. tropicalis, C. guilliermondii and Lodderomyces elongisporus, have a residual content of alkanes which must be removed prior to feeding the yeast to animals. The defatting

Table 11. Lipid composition of a yeast (Candida sp. 107) grown on alkanes (from Thorpe and Ratledge, 1972)

Alkane Substrate	Total lipid (% cell dry wt)	% Lipid composition			PL/CDW (%)
		Sterols*	Triacyl-glycerols	Phospho-lipids	
Dodecane	15	17	56	27	4.1
Tridecane	29	26	51	23	6.7
Tetradecane	26	25	61	14	3.6
Pentadecane	20	27	53	20	4.4
Hexadecane	10	27	53	20	5.0
(Glucose	42	23	72	5	2.1)

* includes sterol esters, free sterols + partial glycerides

process not only removes the unchanged hydrocarbon substrate but also most of the cell lipids. The resulting extract has been termed 'Biolipid' or 'Fermosin' (Voigt et al., 1979). The properties and uses are summarized in Table 12. The harvested yeast is first dried and then extracted with isopropylalcohol/hexane (1:4 v/v) at 60°C. Water is added and the resulting organic phase contains 99.5% of the intracellular (and unused) hydrocarbons and 99.7% of the total lipids of the cell (Grinshpun et al., 1981). The phospholipids were originally described (Voigt et al., 1979) as representing about 25% of the total lipid extract. More recent results (Alenteva et al., 1983) would indicate that the total lipid content can be somewhat reduced in alkane content and that the phospholipid fraction can be up to 49% of the extracted material.

The uses of the Biolipid without fractionation are those as listed in Table 12; most of these uses depend upon the high content of phospholipid. The Biolipid may also be used to prepare individual fractions: ubiquinones, fatty acids, sterols and, of course, phospholipids (Muller, et al., 1982a,b). It is likely though that these applications will be confined to the countries of origin of the Biolipid. No equivalent processes are currently operated outside the USSR and related countries.

OVERALL CONCLUSIONS

The prospects for using microorganisms as potential sources of lipids are limited by the availability of microbial biomass material in sufficient bulk to warrant development of a specific extraction process. Where there has appeared to be a possibility for this, as with bacteria grown on methanol as sources of Single Cell Protein, the processes themselves have not continued in production long enough to encourage long-term interest and investment. The most realistic possibilities probably reside in yeasts being grown in bulk - S. cerevisiae and C. utilis, which are both acceptable as food materials - or in moulds being grown for the production of SCP or

Table 12. Properties and composition of "Biolipid"
(from Voigt et al., 1979)

Origin: East Germany; USSR

Process: Yeasts grown on hydrocarbons and then
defatted.

Composition of oil: Residual alkanes (45-55%) +
microbial lipids: phospholipids
(20-30%), fatty acids (5-10%),
fats (10-20%), sterols (1%).

Appearance: Dark-brown, slightly viscous and
combustible.

Uses: 1. Fuel oil additive to improve dispersion
and flame temperature
2. Mould-release agent in building industry
3. Plant protective agent
4. Surfactant for bitumen
5. Protection layer for deliquescent fertilizers
etc.

Separation of Components: Scheme worked out to allow
purifications of phospholipids (98% pure), etc.

organic acids such as citrate acid or itaconic acid. Some additional possibilities may arise as further interest develops in the use of microorganisms
as sources of lipids in general.

REFERENCES

Aaronson, S., Berner, T. and Dubinsky, Z., 1980, Microalgae as a source of
    chemicals and natural products, in: "Algae Biomass", G. Shelef and
    C.J. Soeder, eds., pp575-601., Elsevier, Amsterdam.
Alenteva, E.S., Velikoslav, D.I. and Bakhmeteva, I.I., 1983, Recovery of
    lipid(s) from yeast biomass - includes thermally ageing thickened
    biomass prior to lipid extraction. USSR Patent 1102276-A.
Alterthum, F. and Rose, A.H., 1973, Osmotic lysis of sphaeroplasts from
    Saccharomyces cerevisiae grown anaerobically in media containing
    different unsaturated fatty acids, J. Gen. Microbiol., 77:371-382.
Ambron, R.T. and Pieringer, R.A., 1973, Phospholipids in microorganisms,
    in: "Form and Function of Phospholipids", G.B. Ansell, J.N. Hawthorne
    and R.M.C. Dawson, eds., Elsevier, Amsterdam.
Anon, 1986, Chemy. Ind., (7 July), p. 437.
Anthony, C., 1982, "The Biochemistry of Methylotrophs", Academic Press,
    London.
Boulton, C.A. and Ratledge, C., 1984, The physiology of hydrocarbon-util-
    izing microorganisms, Top. Enz. Ferment. Technol., 9:11-77.
Boulton, C.A. and Ratledge, C., 1985, Biosynthesis of fatty acids and
    lipids, in: "Comprehensive Biotechnology", vol. 1, M. Moo-Young, ed.,
    pp 459-482, Pergamon Press, Oxford.
Brennan, P.J. and Lösel, D.M., 1978, Physiology of fungal lipids: selected
    topics, Adv. Microbial Physiol., 17:47-179.
Ciferri, O. and Tiboni, O., 1985, The biochemistry and industrial potential
    of Spirulina, Ann. Rev. Microbiol., 39:503-526.
Daum, G., Glatz, H. and Paltauf, F., 1977, Lipid metabolism in an inositol-
    deficient yeast, Saccharomyces carlsbergensis, Biochim. Biophys. Acta,
    488:484-492.
DeBell, R.M. and Jack, R.C., 1975, Stereospecific analysis of major glycero-
    lipids of Phycomyces blakesleeanus sporangiophores and mycelium,
    J. Bacteriol., 124:220-224.
Dubinsky, Z., Berner, T. and Aaronson, S., 1978, Potential of large-scale
    algal culture for biomass and lipid production in arid lands, Biotech.
    Bioeng. Symp. no. 8:51-68.
Farag, R.S., Khalil, F.A., Salem, H. and Ali, L.H.M., 1983, Effects of
    various carbon and nitrogen sources on fungal lipid production.
    J. Amer. Oil Chem. Soc., 60:795-800.
Faust, U., 1979, Process results from SCP-pilot plant based on methanol,
    in: "Microbiology applied to Biotechnology", Dechema Monograph 83;
    pp. 125-134, Verlag Chemie, Weinheim.
Finnerty, W.R., 1978, Physiology and biochemistry of bacterial phospholipid
    metabolism, Adv. Microbial Physiol., 18:177-233.
Fried, A., Tietz, A., Ben-Amotz, A. and Eichenberger, W., 1982, Lipid
    composition of the halotolerant alga, Dunaliella bardawil, Biochim.
    Biophys. Acta, 713:419-426.
Goldberg, I., 1985, "Single cell protein", Springer-Verlag, Berlin.
Goldfine, H., 1972, Comparative aspects of bacterial lipids, Adv. Microbial
    Physiol., 8:1-58.
Goren, M.B. and Brennan, P.J., 1979, Mycobacterial lipids: chemistry and
    biologic activities, in: "Tuberculosis", G.P. Youmans, ed., pp 63-193,
    Saunders, Philadelphia.
Grinshpun, V.V., Zhuchkov, V.N., Volchek, A.M., Neshchadin, A.G. and

Korobiv, Y.K., 1981, Pretreatment of dry biomass by extraction, Ger. (East) Patent 152579 (see Chem. Abs. 1982, 97:4701).

Harwood, J.L. and Russell, N.J., 1984, "Lipids in Plants and Microbes", Allen & Unwin, London.

Henry, S.A., 1982, Membrane lipids of yeast: biochemical and genetic studies, in: "The Molecular Biology of the Yeast Saccharomyces: Metabolism and Expression", J.N. Strathern, E.W. Jones and J.R. Broach, eds., Cold Spring Harbor Laboratory, New York.

Hossack, J.A. and Rose, A.H., 1976, Lipid composition of Saccharomyces cerevisiae enriched with different sterols, J. Bacteriol., 127:67-75.

Hudson, B.J.F. and Karis, I.G., 1974, The lipids of the alga Spirulina, J. Sci. Fd Agric., 25:759-763.

Hunter, K. and Rose, A.H., 1971, Yeast lipids and membranes, in: "The Yeasts", vol. 2, A.H. Rose and J.S. Harrison, eds., pp 211-270, Academic Press, New York.

Hunter, K. and Rose, A.H., 1972, Lipid composition of Saccharomyces cerevisiaé as influenced by growth temperature, Biochim. Biophys. Acta, 260:639-683.

Kaneko, H., Hosohara, M. and Tanaka, M., 1976, Lipid composition of 30 species of yeast, Lipids, 11:837-844.

Lechevalier, M.P., 1982, Lipids in bacterial taxonomy, in: "Handbook of Microbiology", 2nd edn., vol. 4, A.I. Laskin and H.A. Lechevalier, eds., pp. 435-541, CRC Press Inc., Boca Raton, Florida.

Makula, R.A., 1978, Phospholipid composition of methane-utilizing bacteria, J. Bacteriol., 134:771-778.

Mangnall, D. and Getz, G.S., 1973, Phospholipids, in: "Lipids and Biomembranes of Eukaryotic Microorganisms", J.A. Erwin, ed., pp. 145-195, Academic Press, New York.

Mangold, H.K., 1986, Biosyntheses and biotransformation of lipids in plant cell cultures and algae, Chemy. Ind., pp. 260-267.

Materassi, R., Paoletti, C., Balloni, W. and Florenzano, G., 1980, Some consideration on the production of lipid substances by microalgae and cyanobacteria, in: "Algae Biomass", G. Shelef and C.J. Soeder, eds., pp. 619-626, Elsevier, Amsterdam.

Metting, B. and Pyne, J.W., 1986, Biologically active compounds from micro-algae, Enzyme Microb. Technol., 8:386-394.

Milner, H.W., 1951, Possibilities in photosynthetic methods for production of oils and proteins, J. Amer. Oil Chem. Soc., 28:363-367.

Minnikin, D.E. and O'Donnell, A.G., 1984, Actinomycete envelope lipid and peptidoglycan composition, in: "The Biology of the Actinomycetes", M. Goodfellow, M. Mordarski and S.T. Williams, eds., pp. 337-388, Academic Press, London.

Moo-Young, M., 1985, Editor-in-chief "Comprehensive Biotechnology", vol. 2, The Principles of Biotechnology - Engineering Considerations, (with A.E. Humphrey and C.L. Cooney, co-eds.), Pergamon Press, Oxford.

Müller, H., Voigt, B., Riedel, M., Vier, B., Winkler, F. and Hildebrandt, W., 1982a, Verfahren zur Gewinnung von Phosphtiden, Fettsäuren und Ergosterin. Ger. (East) Patent 151007 (see also Chem. Abs. 1982, 96:160821).

Müller, H., Voigt, B., Worbs, M. and Winkler, F., 1982b, Ubiquinones, Ger. (East) Patent 154448 (see also Chem. Abs. 1982, 97:196856).

Nishikawa, Y., Nakanura, I., Kamihara, T. and Fukui, S., 1977, Effects of thiamine and pyridoxine on the lipid composition of Saccharomyces carlsbergensis 4228, Biochim. Biophys. Acta, 486:483-489.

Norris, J.R., 1981, Single cell protein production, in: "Essays in Applied Microbiology", J.R. Norris and M.H. Richmond, eds., pp. 6/1-6/31, John Wiley, Chichester.

Oren, A., Fattom, A., Padan, E. and Tietz, A., 1985, Unsaturated fatty acid composition and biosynthesis in Oscillatoria limnetica and other cyanobacteria, Arch. Microbiol., 141:138-142.

Papacharilaou, E. and Pisano, M.A., 1984, Changes in the lipid composition of Paecilmyces persicinus P-10M1 during growth and cephalosporin C production, Appl. Environ. Microbiol., 48:1084-1087.

Patt, T.E. and Hanson, R.S., 1978, Intracytoplasmic membrane, phospholipid, and sterol content of Methylobacterium organophilum cells grown under different conditions, J. Bacteriol., 134:636-644.

Pieringer, R.A., 1983, Formation of bacterial glycolipids, in: "The Enzymes", 3rd edn., vol. 16, P.D. Boyer, ed., pp. 255-306. Academic Press, New York.

Pohl, P., 1982, Algae lipids, in: "CRC Handbook of Biosolar Resources", A. Mitsui and C.C. Black, eds., vol. 1, pp. 383-404, CRC Press, Boca Raton, Florida.

Prasad, R., 1984, Lipids in the structure and function of yeast membrane, Adv. Lipid Res., 21:187-242.

Prokop, A. and Fekri, M., 1984, Potential of mass algae production in Kuwait, Biotech. Bioeng., 26:1282-1287.

Ramsay, A.M. and Douglas, L.J., 1979, Effects of phosphate limitation of growth on the cell-wall and lipid composition of Saccharomyces cerevisiae, J. Gen. Microbiol., 110:185-191.

Ratledge, C., 1976, The physiology of the mycobacteria, Adv. Microbial Physiol., 13:115-244.

Ratledge, C., 1977, Fermentation substrates, Ann. Rep. Ferment. Proc., 1:49-71.

Ratledge, C., 1980, Microbial lipids derived from hydrocarbons, in: " "Hydrocarbons in Biotechnology", D.E.F. Harrison, I.J. Higgins and R.J. Watkinson, pp. 133-153, Heyden, London.

Ratledge, C., 1982, Microbial oils and fats: an assessment of their commercial potential, Prog. Indust. Microbiol., 16:119-206.

Ratledge, C., 1984, Microbial oils and fats - an overview, in: "Biotechnology for the Oils and Fats Industry", C. Ratledge, P. Dawson and J. Rattray, Amer. Oil Chem. Soc. Monograph no. 11, AOCS, Champaign, Illinois.

Ratledge, C., 1986, The potential of microorganisms for oil production - a review of recent publications, in: "Emerging Technologies in the Oils and Fats Industry",A.R. Baldwin, ed., pp. 318-330, American Oil Chemists' Society, Champaign, Illinois.

Rattray, J.B.M., 1987, Lipids of yeast, in: "Microbial Lipids", vol. 1, C. Ratledge and S.G. Wilkinson, eds., Academic Press, London, in press.

Rattray, J.B.M., Schibeci, A. and Kidby, D.K., 1975, Lipids of yeast, Bacteriol. Rev., 39:197-231.

H.-J. Rehm and G. Reed, 1985, Editors: "Biotechnology - A Comprehensive Treatise", Vol. 2, Fundamentals of Biochemical Engineering. VCH, Weinheim.

Rose, A.H., 1976, (ed.) "Microbial Biomass", Academic Press, London.

Shifrin, N., 1984, Oil from microalgae, in: "Biotechnology for the Oils and Fats Industry", C. Ratledge, P. Dawson and J. Rattray, eds., pp. 145-162, Amer. Oil Chem. Soc., Champaign, Illinois.

Shifrin, N.S. and Chisholm, S.W., 1980, Phytoplankton lipids: environmental influences in production and possible commercial applications, in: "Algae Biomass", G. Shelef and C.J. Solder, eds., pp. 629-645, Elsevier, Amsterdam.

Sinden, K., 1987, Developments in Single Cell Oil production, Enzyme Microb. Technol., in press.

Smith, U. and Ribbons, D.W., 1970, Fine structure of Methanomonas methanoxidans, Arch. Mikrobiol., 74:116-122.

Suzuki, O., Yamashina, T. and Yokochi, T., 1981a, Studies on production of lipids in fungi. I. Lipid compositions of 13 species of deuteromycetes, Yakagaku, 30:854-862.

Suzuki, O., Yakochi, T. and Yamashina, T., 1981b, Studies on production of lipids in fungi. II. Lipid compositions of 6 species of Mucorales in Zygomycetes, Yakagaku, 30:863-868.

Suzuki, O., Yokochi, T. and Yamashina, T., 1982a, Studies on production of lipids in fungi. VI. Changes in lipid composition in three species of the genus Pellicularia of basidiomycetes fungi by cultural conditions, Yukagaku, 31:494-502.

Suzuki, O., Yokochi, T. and Yamashina, T., 1982b, Studies on production of lipids in fungi. VIII. Influence of cultural conditions on lipid compositions of two strains of Mortierella isabellina, Yukagaku, 31:921-931.

Thiele, O.W., 1971, Lipide, Isoprenoide mit Steroiden, G. Thieme Verlag, Stuttgart.

Thomas, D.S., Hossack, J.A. and Rose, A.H., 1978, Plasma-membrane lipid composition and ethanol tolerance in Saccharomyces cerevisiae, Arch. Microbiol., 117:239-245.

Thorpe, R.F. and Ratledge, 1972, Fatty acid distribution in triglycerides of yeasts grown on glucose or alkanes, J. Gen. Microbiol., 72: 151-163.

Tornabene, T.G., Bourne, T.F., Razinddin, S. and Ben-Amotz, A., 1983a, Lipid and polysaccharide constituents of cyanobacterium Spirulina platensis, Marine Ecol. (Prog. Series), 22:121-125.

Tornabene, T.G., Holzer, G., Lien, S. and Burns, N., 1983b, Lipid composition of the nitrogen starved green alga Neochloris oleoabundans, Enzyme Microbiol. Technol., 5:435-440.

Urakami, T., Terao, I. and Nagai, I., 1981, Process for producing bacterial single cell protein from methanol, in: "Microbial Growth on C_1 compounds" H. Dalton, ed., pp. 349-359, Heyden, London.

Vasey, R.B. and Powell, K.A., 1984, Single Cell Protein, Biotechnol. Gen. Eng. Rev., 2:285-311.

Verina, J.N. and Khuller, G.K., 1983, Lipids of actinomycetes, Adv. Lipid Res., 20:257-316.

Voigt, B., Seidel, H., Müller, H., Beck, D., Ringpfeil, M., Reidel, M., Bauch, J., Gentzsch, H. and Bohlmann, D., 1979, Biolipid extrakt- ein neuer Rohstoff ans der Produktion von "fermosin" - Futterhefe auf Basis Erdöldestillat, Chem. Tech. (Leipzig), 31:409-411.

Weaver, T.L., Patrick, M.A. and Dugan, P.R., 1975, Whole cell and membrane lipids of the methylotrophic bacterium Methylosinus trichosporium, J. Bacteriol., 124:602-605.

Weete, J.D., 1980, "Lipid Biochemistry of Fungi and Other Organisms", Plenum Press, New York.

Whitworth, D.A. and Ratledge, C., 1974, Microorganisms as a potential source of oils and fats, Proc. Biochem., 9(9):14-22.

Wilkinson, S.G., 1987, Gram-negative bacterial lipids, in: "Microbial Lipids", C. Ratledge and S.G. Wilkinson, eds., Academic Press, London, in press.

Wood, B.J.B., 1974, Fatty acids and saponifiable lipids., in: "Algal Physiology and Biochemistry", W.D.P. Stewart, ed., pp. 236-265, Blackwells, Oxford.

Yanagimoto, M. and Saitoh, H., 1982, Evaluation tests of a large spiral blue green alga, Oscillatoria sp., for biomass production, J. Ferment. Technol., 60:305-310.

Zepke, H.D., Heinz, E., Radunz, A., Lincheid, M. and Resch, R., 1978. Combination and positional distribution of fatty acids in lipids from blue-green algae, Arch. Microbiol., 119:157-162.

PARTIAL SYNTHESIS OF GLYCEROPHOSPHOLIPIDS

A. Hermetter and F. Paltauf

Institute of Biochemistry and Food Chemistry
Technical University of Graz
Austria

Glycerophospholipids obtained from natural sources, that is from plant, animal or microbial cells, are complex mixtures of phospholipid classes, subclasses and species. It is extremely difficult if not impossible to isolate, with available technology, individual phospholipid species from such mixtures. If one needs a chemically defined phospholipid, e.g., for biochemical or biophysical studies, for the preparation of lipid vesicles or liposomes, or as emulsifiers with well defined properties etc., then chemical synthesis is the only way to make these substances available. The use of synthetic methods is even more compulsory if the desired product is a radioactively or otherwise - spin, fluorescent or photoreactive - labeled phospholipid, or if it contains other "unnatural" rests, such as modified head groups or hydrophobic moieties.

Methods are available to prepare virtually any desired glycerophospholipid or glycerophospholipid-analog by total chemical snythesis. In most cases, however, the procedures described in the literature are quite tedious. For instance, the total chemical synthesis of a mixed-acid unsaturated phosphatidylethanolamine starts from D-mannitol as the chiral educt and comprises 14 reaction steps[1]. If synthesis starts from preformed natural phospholipids, procedures for the preparation of individual glycerophospholipids are considerably shorter. In other words, whenever feasible partial synthesis is much more convenient than total synthesis of complex lipids.

In this report the most widely used semisynthetic procedures for the preparation of glycerophospholipids will be summarized.

In its simplest form, partial synthesis involves only two steps: deacylation of a natural lipid, mostly lecithin, and reacylation with a single fatty acid (Scheme 1). In this way, any defined lecithin with identical short, medium or long-chain acyl rests in both positions can be prepared on any scale, ranging from milligrams up to kilograms.

Similarly, iso-chain phosphatidylethanolamine can be prepared[2], starting from any natural phosphatidylethanolamine that has a low content of ether analogs (Scheme 2). The amino group must be protected during the acylation step in order to avoid formation of the N-acyl derivative. Protection with the phthalyl group has been reported[3], but removal of this protecting group requires treatment with hydrazin, giving rise to the formation of a number of byproducts. The N-trityl group is easily introduced

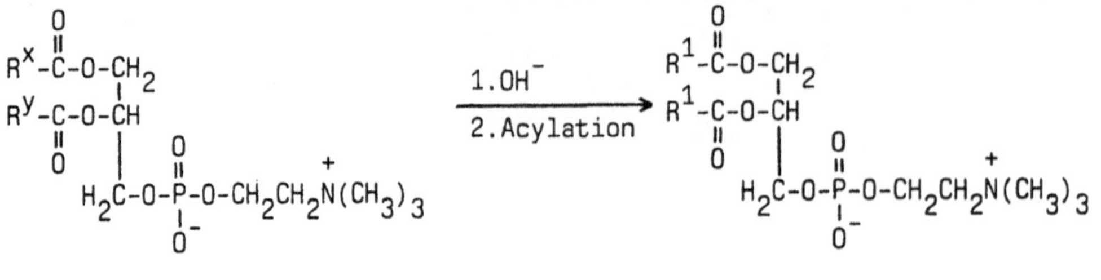

Scheme 1

$$R^x-\overset{\overset{O}{\parallel}}{C}-O-CH_2$$
$$R^y-C-O-\underset{\underset{O}{\parallel}}{CH}$$
$$H_2C-O-\overset{\overset{O}{\parallel}}{\underset{\underset{O^-}{|}}{P}}-O-CH_2CH_2\overset{+}{N}(CH_3)_3$$

$\xrightarrow[\text{2.Acylation}]{\text{1.OH}^-}$

$$R^1-\overset{\overset{O}{\parallel}}{C}-O-CH_2$$
$$R^1-C-O-\underset{\underset{O}{\parallel}}{CH}$$
$$H_2C-O-\overset{\overset{O}{\parallel}}{\underset{\underset{O^-}{|}}{P}}-O-CH_2CH_2\overset{+}{N}(CH_3)_3$$

Scheme 1

by treating phosphatidylethanolamine or a mixture of phosphatidylethanolamine and other phospholipids (e.g., total soya phospholipids) with tritylbromide. Separation of N-trityl phosphatidylethanolamine from other phospholipids is easily achieved. Deacylation and reacylation with the desired
fatty acid poses no problems, nor does deblocking of the amine function.

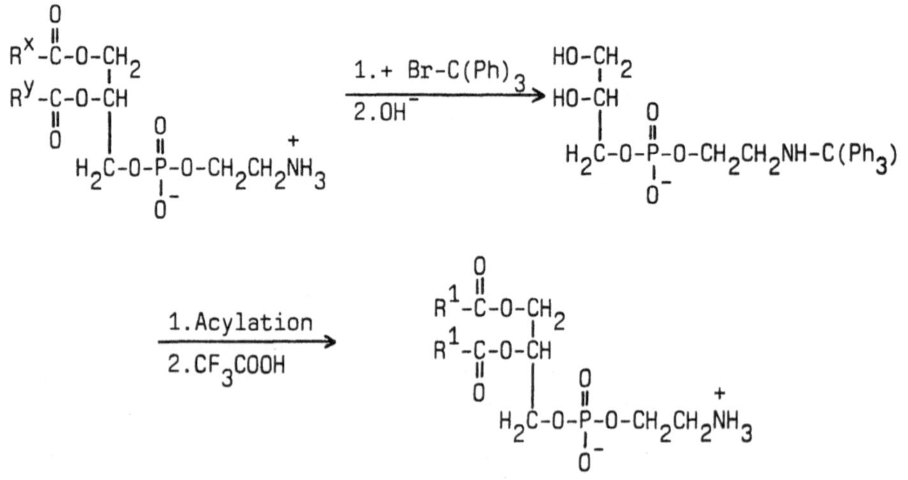

Scheme 2

$$R^x-\overset{\overset{O}{\parallel}}{C}-O-CH_2$$
$$R^y-C-O-\underset{\underset{O}{\parallel}}{CH}$$
$$H_2C-O-\overset{\overset{O}{\parallel}}{\underset{\underset{O^-}{|}}{P}}-O-CH_2CH_2\overset{+}{N}H_3$$

$\xrightarrow[\text{2.OH}^-]{\text{1.}+\ Br-C(Ph)_3}$

$$HO-CH_2$$
$$HO-CH$$
$$H_2C-O-\overset{\overset{O}{\parallel}}{\underset{\underset{O^-}{|}}{P}}-O-CH_2CH_2NH-C(Ph_3)$$

$\xrightarrow[\text{2.CF}_3\text{COOH}]{\text{1.Acylation}}$

$$R^1-\overset{\overset{O}{\parallel}}{C}-O-CH_2$$
$$R^1-C-O-\underset{\underset{O}{\parallel}}{CH}$$
$$H_2C-O-\overset{\overset{O}{\parallel}}{\underset{\underset{O^-}{|}}{P}}-O-CH_2CH_2\overset{+}{N}H_3$$

Scheme 2

Some iso-chain phospholipids are of importance as biologically active
substances, above all dipalmitoyllecithin, which is the major phospholipid
of the lung surfactant. For most other purposes, however, mixed-acid phospholipids are required.

Iso-chain lecithins[4], or N-protected phosphatidylethanolamines[2], are
suitable starting materials for the preparation of mixed-chain derivatives.
The most common procedure involves treatment of the iso-chain phospholipid
with a phospholipase A_2 from snake or bee venom, followed by reacylation
under conditions where acyl-migration is minimal, e.g., with an acyl-anhy-
dride in the presence of dimethylaminopyridine (Scheme 3a). Direct exchange
of the acyl group in position 1 for the desired fatty acid has been achieved
in the presence of Rhizopus delemar lipase[5] (Scheme 3b). This appears to
be an attractive and straightforward procedure if conditions can be worked
out to increase the yield of product. Because of the strict positional
specificity of the lipase involved the product is obtained essentially
free of isomers, whereas in the deacylation-reacylation procedures the
isomer content is between 5 to 10 %.

From the foregoing it is clear that the simple deacylation-reacylation

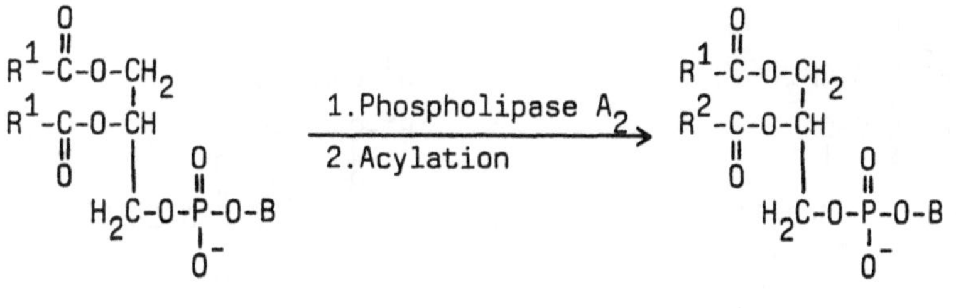

B = choline; N-tritylethanolamine

Scheme 3a

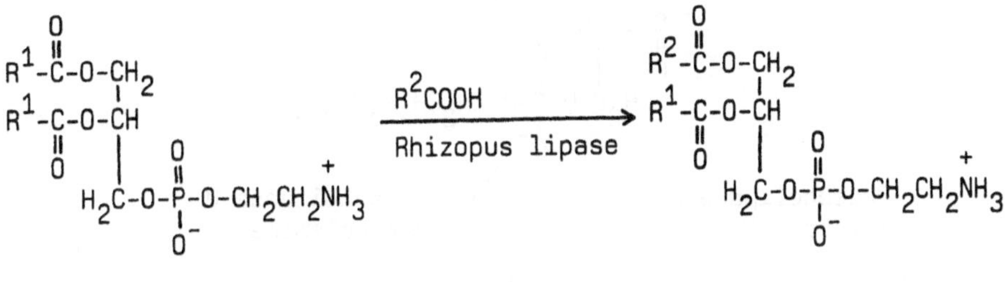

Scheme 3b

procedures for the preparation of iso-chain or mixed-chain phospholipids are easily performed with lecithin and feasible with N-protected phosphatidylethanolamine, but become quite complicated and actually impracticable when the head group contains additional reactive groups as in phosphatidylserine, phosphatidylglycerol or phosphatidylinositol, to name the phospolipids that are quantitatively most important besides phosphatidylcholine and phosphatidylethanolamine. Transesterification of iso- or mixed-chain phosphatidylcholine or phosphatidylethanolamine with the desired head group alcohol (serine, glycerol, choline, ethanolamine, or unnatural primary alcohols) in the presence of a phospholipase D provides an easy access to phospholipids carrying the respective head groups[6] (Scheme 4).

The most widely used phospholipase D is prepared from cabbage. A certain drawback of this enzyme is its selectivity for primary alcohols. Thus, phospholipid derivatives with a secondary alcohol in the head group, e.g., phosphatidylinositol cannot be prepared by this otherwise versatile and elegant method. In a Japanese patent[7] the use of phospholipase DM is recommended for transesterification with a wide variety of alcohols, including secondary and cyclic, aliphatic and aromatic derivatives. This phospholipase DM has been isolated from Nocardiopsis or Actinomadura species.

Phospholipase D- catalyzed transesterification in the headgroup can be performed not only with diacyl analogs, but also with natural or chemically synthesized ether analogs or with otherwise substituted glycerophospholipid derivatives.

Phosphatidic acid, produced from semisynthetically prepared lecithin or phosphatidylethanolamine by phospholipase D- catalyzed hydrolysis can be condensed chemically with the appropriate amino alcohols to produce phosphatidylethanolamines[8,9], phosphatidylcholines[10,11] and phosphatidylserines[12] (Scheme 5).

The positional- and stereo-selectivity of (phospho)lipases has frequently been exploited for the preparation of enantiomeric products from

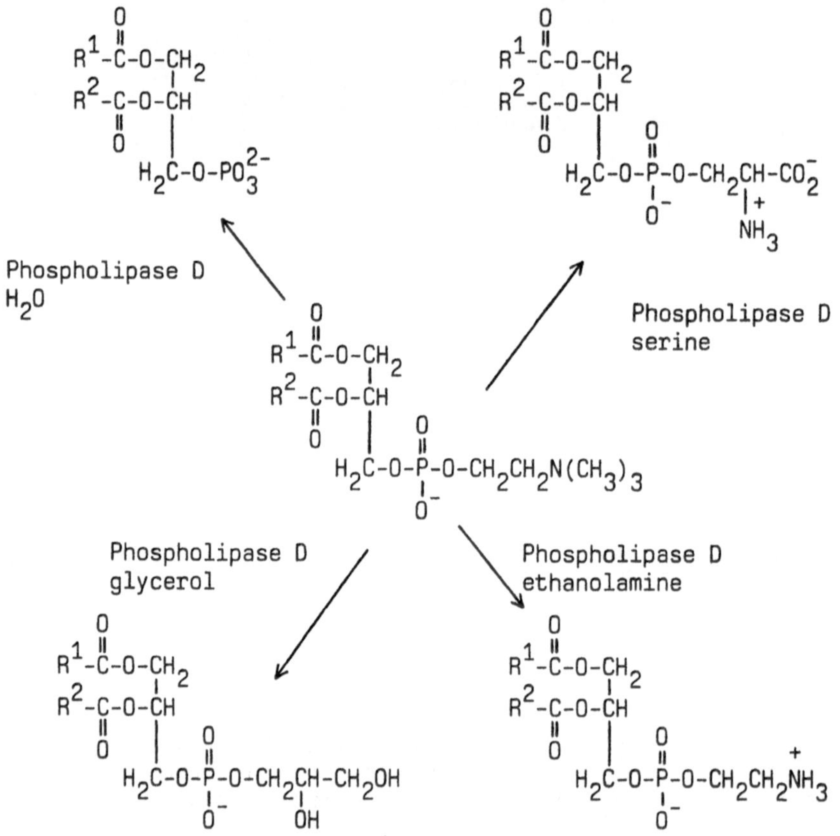

Scheme 4

chemically synthesized racemic educts (c.f., e.g., ref.9). It should be kept in mind that stereoselectivity of phospholipases can be lost if the substrate is modified in critical positions. For example, phospholipases C and D lose stereoselectivity for the sn-3 phospho-enantiomer if in position 2 of the glycerol the natural acyl ester bond is replaced by an ether linkage[13]. Obviously, an ester carbonyl linked to the two-position of glycerol is a prerequisite for stereoselective enzyme action. Similar observations had previously been reported for lipoprotein lipase[14].

Also in semisynthetic procedures the stereoselectivity of enzyme systems can be exploited. An example is given in Scheme 6.

rac.1-O-alkylglycerol, when added to cultured plant cells, e.g. rape or soya, is incorporated into cellular lipids, predominantly into the choline phospholipid fraction with formation of 1-O-alkyl-2-acyl-sn-glycero-3-phosphocholine[15]. This means that only the 1-O-alkyl enantiomer is accepted as a substrate by phospholipid-synthesizing enzymes of the plant cells. Alkylacyl glycerophosphocholine, which is normally absent from these cells, can be hydrolyzed to alkyl glycerophosphocholine which, upon acetylation, yields alkylacetyl glycerophosphocholine, the so-called platelet activating factor (PAF). This ether phospholipid is of utmost interest to cell physiologists because of its extremely high and diverse physiological activity. Numerous procedures have been described for its synthesis. The one mentioned above is quite attractive because it yields a product with a defined alkyl chain. Another semisynthetic approach to PAF starts from choline plasmalogen, isolated, e.g., from beef heart (Scheme 7).

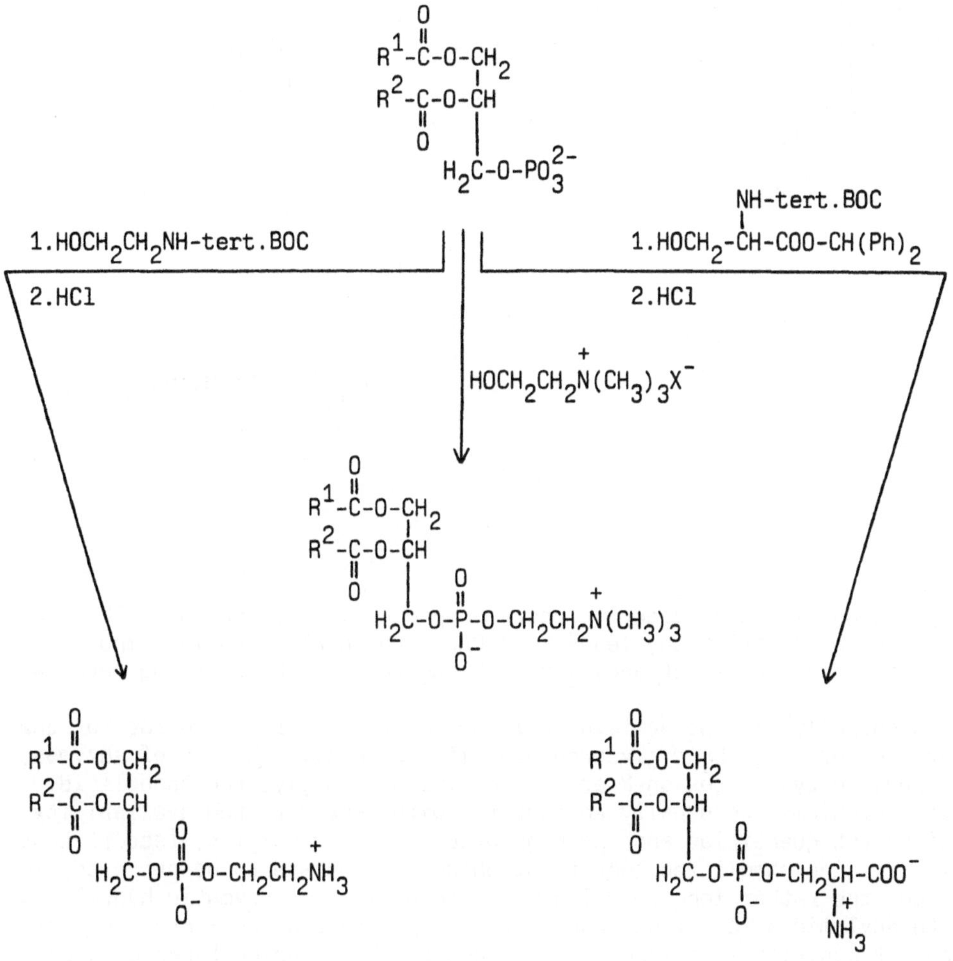

R^1-C-O-CH_2 (O)
R^2-C-O-CH (O)
H_2C-O-PO_3^{2-}

1. HOCH_2CH_2NH-tert.BOC
2. HCl

NH-tert.BOC
1. HOCH_2-CH-COO-CH(Ph)_2
2. HCl

HOCH_2CH_2N^+(CH_3)_3X^-

R^1-C-O-CH_2 (O)
R^2-C-O-CH (O)
H_2C-O-P-O-CH_2CH_2N^+(CH_3)_3 (O, O^-)

R^1-C-O-CH_2 (O)
R^2-C-O-CH (O)
H_2C-O-P-O-CH_2CH_2NH_3^+ (O, O^-)

R^1-C-O-CH_2 (O)
R^2-C-O-CH (O)
H_2C-O-P-O-CH_2CH-COO^- (O, O^-); NH_3^+

Scheme 5

R-O-CH_2
HO···C···H
H_2C-OH

cultured
plant cells
(rape or soya)

R-O-CH_2
R^1-C-O►C◄H (O)
H_2C-O-P-O-CH_2CH_2N^+(CH_3)_3 (O, O^-)

1. OH^-
2. Ac_2O

R-O-CH_2
CH_3-C-O►C◄H (O)
H_2C-O-P-O-CH_2CH_2N^+(CH_3)_3 (O, O^-)

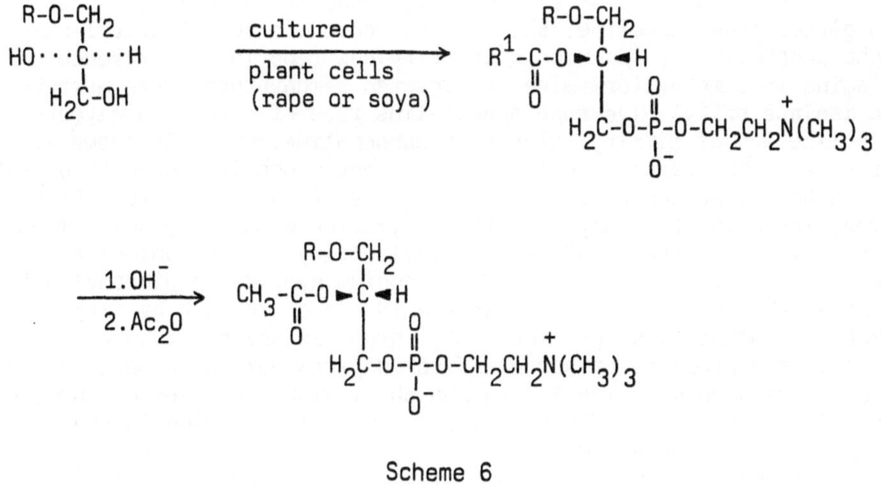

Scheme 6

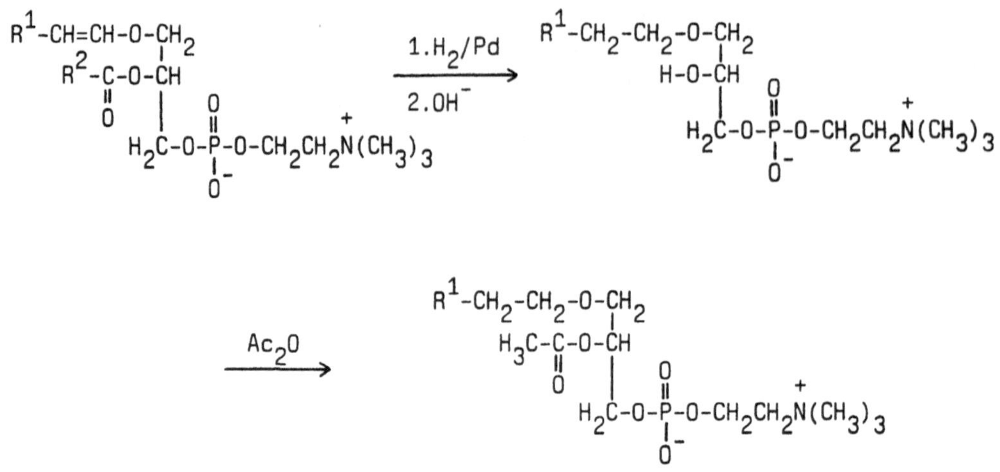

R^1-CH=CH-O-CH$_2$
R^2-C-O-CH
       ‖ | O
       O | ‖
       H$_2$C-O-P-O-CH$_2$CH$_2$N(CH$_3$)$_3$$^+$
             |
             O$^-$

$\xrightarrow[\text{2.OH}^-]{\text{1.H}_2\text{/Pd}}$

R^1-CH$_2$-CH$_2$-O-CH$_2$
H-O-CH
      | O
      | ‖
      H$_2$C-O-P-O-CH$_2$CH$_2$N(CH$_3$)$_3$$^+$
            |
            O$^-$

$\xrightarrow{\text{Ac}_2\text{O}}$

R^1-CH$_2$-CH$_2$-O-CH$_2$
H$_3$C-C-O-CH
       ‖ | O
       O | ‖
       H$_2$C-O-P-O-CH$_2$CH$_2$N(CH$_3$)$_3$$^+$
             |
             O$^-$

Scheme 7

Hydrogenation yields alkylacyl glycerophosphocholine, which upon alkaline hydrolysis and acetylation, leads to a PAF with an alkyl chain composition (mainly hexadecyl and octadecyl) resembling that of the starting material.

Essentially all the semisynthetic procedures described so far for the preparation of mixed-acid glycerophospholipids involve the use of enzymes, with specificity for the sn-2 or sn-1 position of a glycerophospholipid. However, problems are usually encountered with regard to the availability in sufficient quantities and the high price of these enzymes, factors that become limiting when large amounts of phospholipids are to be prepared. In addition, the rather long reaction times required for enzymatic hydrolysis promote acyl migration, thus leading to the presence of an increasing percentage of positional isomers in the products. Considering these problems, we have developed a new concept for the semisynthetic preparation of mixed-chain phospholipids. The new procedure relies only on chemical reactions, thus avoiding the use of expensive enzymes. The method has been worked out for the synthesis of phosphatidylcholines (Scheme 8) and phosphatidyletha-nolamines (Scheme 9). Starting material is the rather inexpensive mixture of soya phospholipids, which contains mainly phosphatidylcholine, phosphat-idylethanolamine and phosphatidylinositol. Phosphatidylcholine is deacy-lated to glycerophosphocholine, and this is converted to 1-O-trityl-sn-gly-cero-3-phosphocholine by reaction with tritylchloride in the presence of triethylamine in dimethylformamide at elevated temperature. Care must be taken to isolate trityl glycerophosphocholine free of unreacted glycero-phosphocholine and of ditrityl glycerophosphocholine, which is formed to some extent when the reaction conditions are not properly chosen. 1-O-trityl glycerophosphocholine can be acylated using any of the common acylation procedures, including the very efficient acylation with fatty acid imida-zolides[16,17] in the presence of the dimethylsulfinylmethide anion; this is obtained by dissolving metallic sodium or sodium hydride in dimethylsulf-oxide. After the reaction, excess fatty acid can be removed by extraction with ammonia in methanol-water. The trityl group is split off under mild conditions in the presence of borontrifluoride/methanol in an aprotic sol-vent at 0°C during 30 min. The 1-lyso phosphatidylcholine need not be iso-lated, but is immediately acylated, e.g., with acyl anhydride in the pre-sence of 4-dimethylamino-pyridine in an aprotic solvent. Preparation of mixed-acid phosphatidylethanolamines follows an analogous procedure[18]. N-trityl glycerophosphoethanolamine is obtained by tritylation of phosphat-

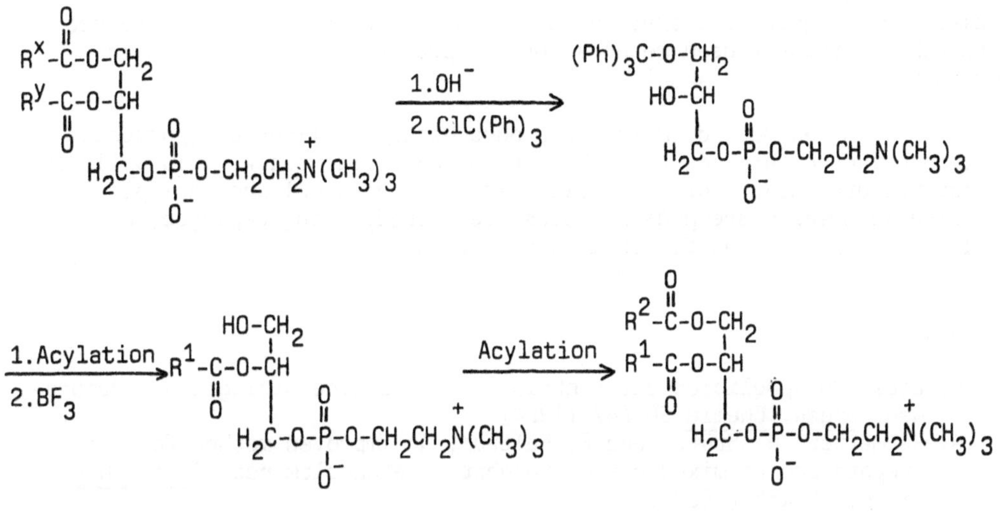

Scheme 8

$$R^x-C(=O)-O-CH_2;\ R^y-C(=O)-O-CH;\ H_2C-O-\overset{O}{\underset{O^-}{P}}-O-CH_2CH_2\overset{+}{N}(CH_3)_3 \quad \xrightarrow[\ 2.ClC(Ph)_3\ ]{\ 1.OH^-\ } \quad (Ph)_3C-O-CH_2;\ HO-CH;\ H_2C-O-\overset{O}{\underset{O^-}{P}}-O-CH_2CH_2\overset{+}{N}(CH_3)_3$$

$$\xrightarrow[\ 2.BF_3\ ]{\ 1.Acylation\ } \quad HO-CH_2;\ R^1-C(=O)-O-CH;\ H_2C-O-\overset{O}{\underset{O^-}{P}}-O-CH_2CH_2\overset{+}{N}(CH_3)_3 \quad \xrightarrow{\ Acylation\ } \quad R^2-C(=O)-O-CH_2;\ R^1-C(=O)-O-CH;\ H_2C-O-\overset{O}{\underset{O^-}{P}}-O-CH_2CH_2\overset{+}{N}(CH_3)_3$$

Scheme 8

idylethanolamine, or a phospholipid mixture enriched in phosphatidylethanolamine (e.g., soya phospholipids), with tritylbromide in the presence
of triethylamine. Base-catalyzed deacylation yields N-trityl glycerophosphoethanolamine and water soluble deacylation products of other glycerophospholipids, which can easily be removed by partition of N-trityl
glycerophosphoethanolamine into a chloroform-methanol phase. O-tritylation
is achieved by standard procedures, using trityl chloride. The resulting
1-O,N-ditrityl glycerophosphoethanolamine is then acylated to yield
1-O,N-ditrityl-2-acyl glycerophosphoethanolamine, from which the 1-O-trityl
protecting group is selectively removed by treatment with borontrifluo-
ride/methanol. Acylation with acyl anhydrides in the presence of dimethylaminopyridine, followed by removal of the N-trityl group with trifluoro-
acetic acid in an aprotic solvent, yields mixed-acid phosphatidylethanolamines in excellent yield. If proper precautions are taken to avoid acyl
migration, the final product contains less than 10 % of the positional

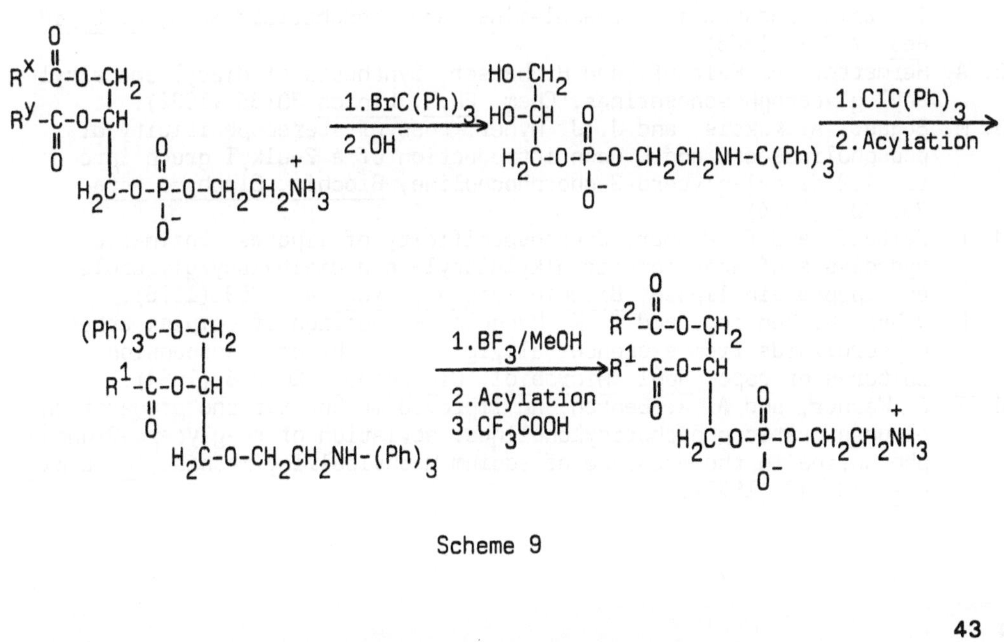

Scheme 9

$$R^x-C(=O)-O-CH_2;\ R^y-C(=O)-O-CH;\ H_2C-O-\overset{O}{\underset{O^-}{P}}-O-CH_2CH_2\overset{+}{N}H_3 \quad \xrightarrow[\ 2.OH^-\ ]{\ 1.BrC(Ph)_3\ } \quad HO-CH_2;\ HO-CH;\ H_2C-O-\overset{O}{\underset{O^-}{P}}-O-CH_2CH_2NH-C(Ph)_3 \quad \xrightarrow[\ 2.Acylation\ ]{\ 1.ClC(Ph)_3\ }$$

$$(Ph)_3C-O-CH_2;\ R^1-C(=O)-O-CH;\ H_2C-O-CH_2CH_2NH-(Ph)_3 \quad \xrightarrow[\begin{array}{c}2.Acylation\\3.CF_3COOH\end{array}]{\ 1.BF_3/MeOH\ } \quad R^2-C(=O)-O-CH_2;\ R^1-C(=O)-O-CH;\ H_2C-O-\overset{O}{\underset{O^-}{P}}-O-CH_2CH_2\overset{+}{N}H_3$$

Scheme 9

isomer. This degree of isomerization is about the same as that obtained with other currently used methods for the preparation of mixed-chain lecithins.

With the new method, mixed-acid phosphatidylcholines and phosphatidylethanolamines are now available at an industrial scale. This will certainly promote the use of chemically defined individual phospholipids in the diverse areas where phospholipids are already being employed, and will open up new fields for their application.

REFERENCES

1. H. Eibl, Phospholipide als funktionelle Bausteine biologischer Membranen, Angew. Chemie 96:247 (1984).
2. A. Hermetter, H. Stütz, and F. Paltauf, An improved method for the preparation of mixed-chain phosphatidylethanolamines, Chem. Phys. Lipids 32:145 (1983).
3. R. Aneja, J. S. Chadha, E. Cubero Robles, and R. Van Daal, Partial synthesis of phosphatidylethanolamines, Biochim. Biophys. Acta 187:439 (1969).
4. C. M. Gupta, R. Radhakrishnan, and H. G. Khorana, Glycerophospholipid synthesis: Improved general method and new analogs containing photoactivable groups, Proc. Natl. Acad. Sci. USA 74:4315 (1977).
5. H. Brockerhoff, P. C. Schmidt, J. W. Fong, and L. J. Tirri, Introduction of labeled fatty acid in position 1 of phosphoglycerides, Lipids 11:421 (1976).
6. P. Comfurius, and R. F. A. Zwaal, The enzymatic synthesis of phosphatidylserine and purification by CM-cellulose column chromatography, Biochim. Biophys. Acta 488:36 (1977).
7. Production of primary or secondary alcohol derivatives of phospholipids by the enzymatic technique, JP 63304/83, and JP 63305/83.
8. I. Barzilay, and Y. Lapidot, The modified synthesis of phosphatidylethanolamine, Chem. Phys. Lipids 3:280 (1969).
9. F. Paltauf, An improved synthesis of 1-O-[^{3}H]alkyl-2-acyl-sn-glycerol-3-phosphoethanolamine with an unsaturated acyl chain, Chem. Phys. Lipids 17:148 (1976).
10. R. Aneja, and J. S. Chadha, A total synthesis of phosphatidylcholines, Biochim. Biophys. Acta 248:455 (1971).
11. A. F. Rosenthal, New, partially hydrolyzable synthetic analogues of lecithin, phosphatidylethanolamine, and phosphatidic acid, J. Lipid Res. 7:779 (1966).
12. A. Hermetter, F. Paltauf, and H. Hauser, Synthesis of diacyl and alkylacyl glycerophosphoserines, Chem. Phys. Lipids 30:35 (1982).
13. M. Bugaut, A. Kuksis, and J. J. Myher, Loss of stereospecificity of phospholipases C and D upon introduction of a 2-alkyl group into rac-1,2-diacyl-glycero-3-phosphocholine, Biochim. Biophys. Acta 835:304 (1985).
14. F. Paltauf, and E. Wagner, Stereospecificity of lipases. Enzymatic hydrolysis of enantiomeric alkyldiacyl- and dialkylacylglycerols by lipoprotein lipase, Biochim. Biophys. Acta 431:359 (1976).
15. N. Weber, H. Benning, and H. K. Mangold, Production of complex ether glycerolipids from exogenous alkylglycerols by cell suspension cultures of rape, Appl. Microbiol. Biotechnol. 20:238 (1984).
16. T. G. Warner, and A. A. Benson, An improved method for the preparation of unsaturated phosphatidylcholines: acylation of sn-glycero-3-phosphocholine in the presence of sodium methylsulfinylmethide, J. Lipid Res. 18:548 (1977).

17. A. Hermetter, and F. Paltauf, A facile procedure for the synthesis of saturated phosphatidylcholines, Chem. Phys. Lipids 28:111 (1981).
18. A. Hermetter, H. Stütz, K. Lohner, and F. Paltauf, 1-O,N-Ditritylgly-cerophosphoethanolamine, a novel intermediate for the facile preparation of mixed-acid phosphatidylethanolamines, Chem. Phys. Lipids, in press.

DRUG ENTRAPMENT BY PHOSPHOLIPIDS

Rolf E. Schubert and Karl-Heinz Schmidt

Department of Surgery
University of Tuebingen
West Germany

SUMMARY

In living organisms, metabolically active compounds are frequently
associated with aggregated phospholipid structures. For example, hydrophilic
molecules (e.g., enzymes, hormones, neurotransmitters) can be found inside
the cell as the content of phospholipid vesicles. In addition, hydrophobic
biomolecules like triglycerides or cholesterol are transported in the form
of aggregates such as mixed micelles or lipoproteins. This entrapment of
hydrophilic and hydrophobic biomolecules is possible due to monolayer formation on hydrophobic surfaces and bilayer formation on hydrophilic surfaces.

As a consequence of the evolution of various phospholipid species,
different parameters of the lipid interface (e.g., surface charge and hydration, internal order, viscosity, conformation) can be modified. The inter-
facial properties of the phospholipid aggregates determine their fate and
consequently the distribution pattern of the entrapped molecules.

The technology of drug entrapment by phospholipids, therefore, follows
well established biological models and, in essence, yields predictable results on stability and release properties, adhesion, transport, endocytosis,
and fusion.

INTRODUCTION

In nature, the primary function of phospholipids (PLs) is to form
membranes. Membranes can separate compartments, as well as store or
transport substances. Vesicles can be separated from specific membrane
areas, which, in turn, can fuse with other vesicles or membranes.

Inside the cell, the Golgi apparatus segregates small vesicles with
distinct surface structures. This segregation takes place during the
processing of molecules that are later released from the cell or delivered
into the plasma membrane after fusion.

In the vascular system, erythrocytes are large spezialized transport
vesicles with exceptional membrane properties. On the one hand, in spite of
their size, they are deformable enough to pass through capillaries and
sometimes even endothelial gaps. On the other, their membranes are not
significantly destabilized by plasma constituents and, as long as their
elasticity and membrane asymmetry are within a defined range, they are not
attacked by macrophages (McEvoy et al., 1986).

Vesicular systems primarily entrap hydrophilic molecules. PLs, however, are also capable of binding or entrapping amphiphilic or lipophilic molecules.

In bile, together with bile salts and cholesterol, PLs form mixed micelles. These aggregates solubilize the hydrophobic cholesterol for biliary excretion, and drastically reduce the free portion of the large amount of bile salts and thus helping to avoid damage by these natural detergents to cell membranes.

In blood and lymph, plasma lipoproteins are transport vehicles for cholesterol and triglycerides to peripheral regions. In this case, PLs, by forming a stable emulsion, pack strongly hydrophobic substances in a hydrophilic environment.

In natural systems, PLs do not necessarily form bilayers or closed monolayers to entrap substances, and there are probably some poorly characterized PL aggregates. In many cases, PLs require specific proteins to perform their functions. Nevertheless, only the variability in their hydrophobic regions and the evolution of many different species, the hydrophilic headgroups of which differ, enable PL aggregates to adopt a variety of conformations. The aggregates, therefore, can be stable or undergo reactions such as deformation, fusion, endocytosis or exocytosis, lipid exchange, and delivery or binding of molecules.

Only precise knowledge of the conditions of these reactions enables optimal biomimetical entrapment of drugs into PL aggregates and their delivery at the target.

PHOSPHOLIPIDS AND THEIR AGGREGATION BEHAVIOR

The polymorphic phase properties of PLs are important for their consideration as drug carriers. Depending on lipid mixture, influence of entrapped drugs, interactions with the organism, temperature during storage and in vivo, PL aggregates can adopt different conformations. Cullis et al. (1985) excellently reviewed the in vitro situation of different PL species and mixtures; increasing attention, however, is now being directed to the more complex in vivo situation.

In short, when PL monomers form aggregates, they can adopt, with a certain degree of probability, micellar (spherical), hexagonal (cylindrical), lamellar, or three-dimensional conformations with high curvatures that can be summarized as lipidic particles (LIPs). These structures have different energy levels and phase transitions occur at definite transition temperatures or when counterions or hydration change. Structures and transitions are detectable by, for example, X-ray diffraction, NMR, ESR, freeze fracture electron microscopy, and calorimetry. The most important transition for membranes is that from the lamellar gel (L_β) to the lamellar liquid crystal phase (L_α); it occurs at a more or less narrow range of temperature T_m. Moreover, non rotating PLs, tilted chains (L_β') of PL aggregates, rippled membrane structures (P_β), and even SUV, LUV, and MLV (see below) may be considered distinct phases in lamellar systems. Hexagonal structures, i.e., tubes with hydrophilic headgroups inside (H_{II}) or rods with hydrophilic headgroups outside (H_I), and LIPs such as cubic phases, inverted micelles, or bulged bilayers have been found in artificial systems. Presumably, they are intermediates in natural membranes formed during membrane fusion and other destabilized conditions.

Many properties of drug-carrying PL aggregates such as stability or inducibility of fusion, therefore, depend on the choice of the appropriate PL species.

PL species differ in their headgroups, which are additionally influenced by the degree of hydration, charge and counterions; neighboring lipids allow varying degrees of hydrogen bonds. In the hydrophobic part, the mean shape is determined by length, degree of saturation, branching, number of acyl chains and motion. Consequently, headgroup area and hydrocarbon extension determine the shape of a single membrane lipid molecule, and the sum of the shape properties in aggregates determines the phase properties adopted. (Israelachvili et al., 1980).

In general, cylindrical PLs forming bilayers are the neutral phosphatidylcholine and sphingomyelin, the negatively charged phosphatidylserine, phosphatidylglycerol, phosphatidylinositol, phosphatidic acid and cardiolipin, as well as some glycolipids.

Micellar structures are preferred by lipids with inverted cone shape (e.g., detergents and lyso-phospholipids), whereas cone shaped PLs form hexagonal (H_{II}) structures (e.g., phosphatidylethanolamine at elevated temperatures, and negatively charged PLs at low pH or in the presence of some divalent cations).

It, however, should be noted that entrapped drugs or attached plasma constituents, through their interaction with hydrophobic and/or hydrophilic regions, may change the phase properties and, consequently, the binding or barrier characteristics of PL aggregates. It, therefore, is not surprizing that, on the one hand, PLs such as lyso-phospholipids that tend to form H_I structures or micelles are stabilized to bilayers by drugs like gramicidin that intercalate among hydrophobic chains (Killian et al., 1983) and, on the other, phosphatidylethanolamine molecules that form H_{II} structures are stabilized to bilayers by molecules (e.g., detergents) which insert into the head group region (Madden and Cullis, 1982).

Knowledge of the drug/phospholipid interaction may replace the use of standard lipid mixtures or occult receipes. There is virtually an inexhaustible source of different PLs in nature, even for lipids with exceptional properties like those from archebacters, that protect their cell against osmotic rupture and chemical attack (Ring et al., 1986). In addition, several strategies for chemical synthesis of PLs or other amphiphiles (Eibl, 1981) increase the possibility of using tailored membrane lipids.

PHOSPHOLIPID AGGREGATES

The nature of PL aggregates is determined by the chemical structure and environment of the PL. In hydrophilic solvents, formation of flat aggregates with edges is highly unlikely with PLs capable of forming bilayers since high edge tensions lead to closure of the bilayers to vesicles. A sufficient amount of detergent molecules such as bile salts may reduce the edge tension (Fromherz et al., 1986) and, therefore, stabilize membrane sheets as mixed micelles. Large quantities of strongly hydrophobic substances, with parallel intercalation between the hydrocarbon chains and destabilization of the membrane bi- or monolayer, cannot be entrapped by PLs. The better they form a hydrophobic domain which does not disturb the acyl chain region of the PLs, the better they can be entrapped as a hydrophobic core coated by a PL monolayer.

Plasma lipoproteins, mixed micelles and vesicles are the most common PL-containing aggregates.

Lipoproteins

The different plasma proteins share some common characteristics. Amphipathic phospholipids and cholesterol, together with the apoproteins, cover the surface and are well suited for entrapping hydrophobic lipids like triglycerides and esterified cholesterol in the center of the aggregate. The phospholipid cover to form a monolayer that makes the inner apolar core inaccessible to extracellular enzymes is progressively closer with decreasing size of the lipoproteins. Since the not catalyzed exchange of cholesteryl esters and triglycerides of low density lipoproteins (LDL) with other membranes is negligible, they must be internalized into the cell before the hydrophobic lipids can be utilized for metabolic or structural purposes. During this process, binding of LDL to a cell receptor must occur before endocytosis can be initiated.

Moreover, lipoproteins must be small enough to penetrate through the gaps between the endothelial cells of the arterial wall. High density lipoproteins (HDL, with a diameter of 6-14 nm) and LDL (15-25 nm), therefore, can reach peripheral tissue cells, whereas very low density lipoproteins (VLDL, 30-70 nm) and chylomicrons (100-1000 nm) have to be catabolized in the blood.

An emulsion of hydrophobic substances as core entrapped in a phospholipid monolayer cover must satisfy certain requirements, if these natural structures are to be followed biomimetically. Specific size, together with certain surface proteins, enable a provided tissue distribution. Endocytosis of these aggregates by specific tissue cells is possible only when these artificial aggregates are small enough, since the tightly packed hydrophobic drugs in the core may hamper the deformability of the aggregate necessary for passage through capillaries and the endothelial gaps.

Mixed Micelles

Detergents are capable of binding hydrophobic molecules like lipids to form mixed micelles. Bile acids, i.e., natural detergents, are able to solubilize dietary fats in the gut, and, together with lecithin, cholesterol in the bile. The structure and behavior of these mixed micelles in vitro is well known. At a bile salt/phospholipid ratio below 2:1 (Müller, 1981), the aggregates form mixed disk micelles, i.e., diskoidal bilayer fragments of membrane lipids and bile salts. These detergents can be inserted between the lipid molecules (Mazer et al., 1976), but primarily form the outer edge of the disk (Small, 1971). At high bile salt content the shape of the mixed micelles is transformed into smaller ellipsoids, the actual shape in the bile.

Some applications of mixed micelles as carriers of hydrophobic drugs appear to be successful. For example, mixed micelles containing diazepam are apparently tolerated in the organism. One advantage of mixed micelle application, in addition to the entrapment of hydrophobic drugs, could be the small size of the aggregates (approx. 10,000 daltons and larger); in any case, these aggregates are small enough to pass through the gaps between the endothelial cells of the vascular system.

Liposomes

Natural membranes usually have a bilayer structure that separates hydrophilic compartments and also allows biopolymers to anchor. Membranes are able to adopt low or high curvatures, depending on their functions as larger membranes, protrusions, or vesicles. Intracellular vesicles have a diameter of approximately 100-200 nm. Human erythrocytes, i.e., intravascular vesicles, in spite of their size, are highly flexible. In membranes, the concerted action of different membrane lipids and proteins is now better understood. Lipids are also unequally distributed between the two monolayers, and connections to the cytoskeleton network enable the lipid domains to undergo local bilayer destabilizations without total membrane breakdown. Attempts have been made to prepare suited artificial membranes by mimicking some of these membrane capacities (e.g., storage, binding, fusion).

Liposomes are artificial vesicles, which the New York Academy of Sciences (1977) has defined as "multilamellar large vesicles (MLVs), size range 0.1-5 μm; small unilamellar vesicles (SUVs), size range 0.02-0.05 μm; and large unilamellar vesicles (LUVs), size ranging from 0.06 μm". This distinction points to essentially three different aggregate structures and their varied suitability as drug carriers.

MLVs have concentric lipid bilayers with an onionskin configuration. They are easily prepared and were the first liposomes studied (Bangham et al., 1965). Since these early studies on their barrier characteristics, they have proven valuable, for example, for fundamental research on membrane conformations. Due to their size and lack of similarity to natural membranes, modified MLVs have been used to activate macrophages after phagocytosis (Fidler and Poste, 1982).

SUVs are also easily prepared from MLVs and were helpful in clarifying the principles of PL membrane structures (Huang and Mason, 1978). Their high curvature makes them behave differently from relaxed membranes. For example, in small egg yolk lecithin vesicles, the inner monolayer is thinner (16 Å) than the outer monolayer (21 Å). Headgroups and tails are compressed differently in the two monolayers, and the volume of the outer monolayer is approximately twice that of the inner layer. Finally, since the ratio of hydrated lipid volume to trapping volume is greater than 5, SUVs are not suitable for satisfactory entrapment of hydrophilic drugs. They, however, may be usable with more hydrophobic drugs that bind to the hydrocarbon part of the aggregates. In this case, the trapping efficiency depends more on the amount of PL than on vesicle size. Even drugs that have increased hydrophobicity with decreased proton concentration may be well bound to the inner monolayer; at pH = 7, only 1 in approximately 10 of these vesicles contains a free trapped proton (Cullis et al., 1985). Another advantage of SUVs may be their small size, when the target is a peripheral tissue cell. As described above, aggregates smaller than 25 nm can pass through the endothelial gaps of the vascular system.

LUVs have an advantageous trapping volume. Different preparation techniques admit a large size range with low polydispersity (i.e., high homogeneity). This, together with their similarity to natural bilayers, are the main advantages, and researchers publishing new liposome preparation methods, therefore, frequently report having obtained almost exclusively unilamellar structures.

Most liposomes used for fundamental research in drug entrapment can be regarded as liposomes of the first generation. Two differences from biological membranes are striking: the lack of stability and the almost symmetric distribution of lipid species between the two monolayers. Attempts have been made to improve artificial vesicles. Cholesterol, indeed, enhances membrane order without necessarily reducing membrane fluidity, but the most important advancement seems to be the lateral polymerization of membrane PLs and, consequently, the mimicking of the stabilization of natural membranes by the cytoskeleton (Hayward et al., 1985; Regen, 1985; Bader and Ringsdorf, 1986; Chapman et al., 1986). The elasticity behavior of polymer vesicles is similar to that of natural membranes, as demonstrated recently by comparing the shapes of erythrocytes and headgroup-polymerized vesicles (Sackmann et al., 1986).

Lipid asymmetry in liposomes is still considered to be of less importance, but it could help solving stability problems of liposomes in the blood, as will be shown below. Erythrocytes are apparently eliminated by macrophages, when their flexibility and lipid asymmetry is disturbed (McEvoy et al., 1986). This should be kept in mind, when LUVs with enhanced lifetimes in the vascular system are prepared.

PREPARATION METHODS

Emulsions of hydrophobic drugs and PLs can be formed mechanically. Mixed micelles are easily prepared by dissolving a dry mixture of hydrophobic drugs, PLs and detergents in hydrophilic solvents. Liposomes can be formed by a variety of methods. The listing presented here, while not definitive, nevertheless, differenciates five main groups:

Mechanical Procedures

Dried lipids spontaneously form MLV when buffer is added (Bangham et al., 1965). Thereafter, MLV can be sized by extrusion through polycarbonate membranes to obtain smaller MLV (Olson et al., 1979) or LUV (Hope et al., 1985). MLV can also be sized by sonication (Huang, 1969), using a french press (Barenholz et al., 1979) or a microemulsifier (Mayhew et al., 1984).

Use of Organic Solvents

Injection of lipids, dissolved in organic solvents like ethanol (Batzri and Korn, 1973) or ether (Deamer and Bangham, 1976), into buffer solution results in the formation of SUV or LUV, respectively, while organic solvent is diluted or vaporizes. Another possibility is the mild sonication of lipids and buffer in organic solvents. This procedure leads to the formation of relatively stable emulsions of inverted micelles (prevesicles) with buffer entrapped in spheres of PL monolayers with surrounding organic solvent. Thereafter, solvent is removed by evaporation ("reverse phase evaporation") and liposomes are formed at a critical solvent content (Szoka and Papahadjopoulos, 1978). Prevesicles may also be mixed with other lipids and, subsequently, centrifuged through a lipid monolayer into buffer. Lipid asymmetry in vesicles can be achieved by this method (Träuble and Grell, 1971).

Use of Detergents

Detergents in mixed micelles are in equilibrium with detergent mono-mers. Removal of these monomers reduces the detergents bound to mixed micelles. This, in turn, forces lateral fusion of the aggregates and finally the formation of vesicles. Monomer removal can be performed by column chromatography (Brunner et al., 1976), dialysis using hollow fibers (Rhoden and Goldin, 1979) or polycarbonate membranes (Milsmann et al., 1978), or by dilution (Schurtenberger et al., 1984). Vesicle size depends mainly on the detergent used, but also on the temperature and the rate of detergent removal.

Electroformation

This method, recently reported by Angelova and Dimitrov (1986), is a gentle procedure, which may prove suitable for large scale preparation.

Subsequent Modifications

After preparation, liposomes can be modified. For example, negatively charged membranes can be fused with Ca   and, thereby, vesicle size can be enhanced (Papahadjopoulos et al., 1975). Different species of molecules can be coupled to membrane surfaces (Gregoriadis, 1984). Moreover, new concepts of mimicking natural membrane stability are the lateral polymerization (Regen, 1985; Bader and Ringsdorf, 1986; Chapman et al., 1986) of natural or synthetic lipids, as well as the preparation of asymmetric vesicles. Since the early findings of Wirtz (1974) that specific proteins are capable of exchanging lipids from the outer monolayers of membranes, these lipid trans-fer proteins from different sources may become a useful tool for achieving lipid asymmetry.

There is no one ideal method for preparing liposomes. The selection of the method depends on the aim of the liposome application. The main aspects to consider are the irritability of the drug to be entrapped and the mem-brane components (including proteins) constituting the trap. Other important factors to be taken into consideration are trapping efficiency, size of li-posomes, their unilamellarity or oligolamellarity, and residues of solvents or detergents.

DRUG ENTRAPMENT

Drug entrapment can be performed during preparation of PL aggregates by adding hydrophilic drugs to the preparation buffer or strongly hydrophobic drugs to the lipids. Moreover, there are several ways of entrapping drugs after liposome preparation.

Freezing and thawing (F/T), which, due to hydration changes, temporarily changes membrane conformations, is suitable in some cases. Hydrophobic drugs tending to diffuse across the membrane can be entrapped, if they bind to chelating agents that were entrapped during liposome preparation. For less diffusable drugs, membranes can be permeabilized temporarily by suitable agents.

Afterwards, not entrapped drugs can easily be separated from liposomes by dialysis, gel chromatography, or rapid centrifugation of the suspension through chromatography gel (Lelkes, 1984), resulting in negligible dilution.

Since the first studies on the diffusion behavior of ions across artificial membranes (Bangham et al., 1965), successful and unsuccessful attempts have been made to entrap many different substances in liposomes and other PL aggregates. To list of these substances would go beyond the scope of this presentation. It, however, should be borne in mind that drugs must frequently be chemically modified (prodrugs) to before the trapping efficiency of PL aggregates can be enhanced (Schwendener et al., 1986).

IN VITRO STABILITY

The practicability of storage depends on the membranes used and the drugs entrapped. There are no general rules. Investigations of liposomal suspensions after storage, freezing, freeze drying and freezing/thawing cycles show different results for different lipid/drug combinations (Crommelin and van Bommel, 1984; Strauss, 1984; Stricker and Kibat, 1986). In most cases, however, liposomes are not suited for long-time storage and the best quality is obtained with freshly prepared liposomes. On the one hand, lipid oxidation cannot be completely excluded, in spite of antioxidants in the membrane. On the other, amphiphilic drugs are easily released, thus reducing the entrapped portion.

Liposomes, therefore, should be prepared as late as possible before application.

STERILITY

In general, microorganisms are destroyed by heat or irradiation, chemical attack via toxic gases or antibiotics, or they are separated by filtration or centrifugation.

In the case of PL aggregates, heat or irradiation are less suited, because they may cause oxidizing or polymerizing reactions of the aggregate constituents. Gas sterilization of ready preparations may not solve the problem, because gases dissolve well in a hydrophobic environment and, therefore, remain as toxic residues in the lipids or hydrophobic drugs. Even antibiotics, if they are not the drugs to be entrapped, must frequently be avoided.

Materials and equipments, therefore, should be sterile at the beginning of PL aggregate preparation. If this is not the case, ready preparations must be separated from contaminates by filtration, adsorption, or centrifugation, using different size, surface, or density of PL aggregates and microorganisms.

APPLICATIONS

Therapies

In addition to orally applied natural PL aggregates for nutritional purposes, PL emulsions are probably the most frequently applied PL aggregates; they are used clinically in the parenteral nutrition regimens together with nutritional fats. The number of different application areas to date, however, is low, approaching that of mixed micelles.

By contrast, the possibilities of liposome applications are extensive. Drugs encapsulated by liposomes are used in therapy for cancer (for reviews, see Weinstein and Lesermann, 1984; Coune, 1984) and arthritis (Knight et al., 1985), as well as in the fields of dermatology (Westerhof, 1985) and ophthalmology (Lee et al.,1983). Liposomes are useful in immunology (Alving and Richards, 1983; Sullivan et al.,1986), for correction of enzyme defects (Patel and Ryman, 1981), for detoxification procedures (Wendel et al., 1982), and, as blood substitute with entrapped hemoglobin (Hunt and Burnette, 1984; Hayward et al., 1985; Pirkl et al., 1986).

In addition to in vivo applications, liposomes are also employed in molecular biology (Ostro and Giacomoni, 1983), cell research (Huang, 1983), and in industry, e.g. for separation of racemic drugs by using chiral elements in liposomal membranes (Roth, 1986).

Routes

Depending on the difficulties and possibilities of reaching target tissues and in addition to the applications described above, liposomes have been applied on different in vivo sites that are reviewed elsewhere (Patel and Ryman, 1981; Mayhew and Papahadjopoulos, 1983). Liposomes can also be applied endolymphatically (Hirnle et al., 1986), intramuscularly (Arrowsmith et al., 1984), and per rectum (Gabev et al., 1985).

IN VIVO BEHAVIOR

The first aspect to be considered with respect to in vivo behavior is, whether or not PL aggregates themselves are toxic.

PL emulsions lead to embolism, when the amount of fat is too high or the particles are too large.

Cooperative binding of bile salts to mixed micelles together with hampered bile salt binding to membranes (Schubert et al., 1986a) could explain the coexistence of mixed micelles and phospholipid vesicles (and, therefore, the intactness of natural membranes such as the canalicular part of the liver cell plasma membrane) even at high bile salt concentrations of approximately 10 mmol/l, as demonstrated in bile specimen (Sömjen and Gilat, 1985).

It should be noted, however, that bile salt binding equilibrium leads to removal of detergent molecules from mixed micelles during dilution, leading to reactive phospholipid aggregates. In in vitro systems, the reaction leads to the formation of vesicles (Schurtenberger et al., 1984). Consequently, many in vivo reactions of diluted mixed micelles are possible.

Liposomes of suitable size, which can bypass hinderances, interact with cells only because of their specific membrane properties. The primary interactions with cells in vitro are lipid exchange, fusion, stable adsorption, or endocytosis (Pagano et al., 1981); these interactions may also occur in vivo.

By interaction with tissue cells, PL aggregates can induce metabolic reactions or releases of cell constituents, which ultimately could harm these cells or the organism. Apparently, even large volumes of intravenously administered neutral liposomes (400 ml, 8 g lipids) are well tolerated by humans (Coune et al. 1983). The risks of administering fat emulsions, therefore, cannot compared with those of liposomes. Certain lipids that are frequently used to induce liposomal surface charge (e.g., stearylamine or dicetylphosphate) and even phosphatidylserine should be avoided, as shown in cell culture experiments (Campbell, 1983). In addition, the metabolic pathways of exotic membrane lipids like archebacterial lipids or lipid polymers have not yet been clarified.

The next aspect to consider is the in vivo stability of PL aggregates. Scherphof et al. (1981, 1983, 1984) demonstrated and reviewed some liposome alterations by several plasma constituents. In short, liposomes can be destabilized, especially by high density lipoproteins (HDL) and probably by the complement system, if insufficient quantities of cholesterol, sphingo-myelin, or glycolipids are present in their membrane. Liposomal membrane damage may vary from increased permeability to complete destruction.

Recently, we demonstrated in a model system using detergents that, in many cases, increase of permeability in liposomal membranes is not only due to an increase in membrane fluidity, but it is also primarily a short-time phenomenon which occurs during a membrane foldover (Schubert et al., 1986b). This foldover is induced by membrane stress in the outer monolayer following binding of external molecules. After resealing of transient membrane holes, membrane stability is almost totally restored. Should this mechanism be valid under in vivo situations of liposomes as well, polymerization and even lipid asymmetry might help protect membrane stability, providing liposomal membrane order in the outer monolayer is lower than in the inner monolayer.

The paucity of knowledge on physiologic and metabolic pathways of mixed micelles and PL emulsions as well as their possible reactions in the organism may be due to a lack of interest in these PL aggregates.

TARGETING

All successful attempts to prepare PL aggregates that are capable of reaching and acting exclusively on the target tissue or cell, are referred to as targeting.

Targeting is successful only when drug entrapment in PL aggregates and their stability is sufficient and after aggregates have reached their target unaffectedly. In most cases, however, this has only been possible in the in vitro targeting of PL aggregates to cells (Pagano et al., 1981). Procedures to optimize specific interactions with target tissue cells in vivo are the most difficult and, therefore, were left to last. As mentioned above, targeting can be partially achieved by selecting the size of the aggregates and by their application route in the organism. Once the PL aggregate reaches the target cell, it must be attached before fusion or internalization. If these single reaction steps are to occur specifically with a certain population of cells, the surface structures of the aggregate must be adapted to that of the cell. This represents a large immunological field for specialists in protein and sugar research. Some modification procedures for liposomes have been reviewed elsewhere (Gregoriadis, 1984). In special cases, however, in vivo fusion with target cells can be achieved by the choice of suitable membrane lipids without the use of proteins or modifications (Cevc et al., 1986).

CONCLUSIONS

Drug entrapment by PL aggregates is a useful technique in pharmaco-therapy for improving pharmacokinetics, tissue distribution and bioavail-ability of drugs. Moreover, toxicity and other potential side effects of drugs can be controlled by application of this technique. In particular, the handling of hydrophobic drugs has been significantly improved.

In biomimetic approaches, the technology of drug entrapment by PL aggregates follows the pathways of the cellular processing of metabolically active tissue components like hormones, neurotransmitters or enzymes. In the future, advances in the efficacy of pharmacotherapy can be increased by improving the specific targeting of the PL entrapped drug and the controlled interaction with the target.

Acknowledgments

The authors would like to express their appreciation to Mr. Rainer Storf for his help in the preparation of the manuscript and to Ms. Loni Schweikert and Mr. Jürgen Haab for the helpful advice.

REFERENCES

Alving, C. R., and Richards, R. L., 1983, Immunologic aspects of liposomes, in: "Liposomes", M. J. Ostro, ed., Marcel Decker, New York - Basel.

Angelova, M. I., and Dimitrov, D. S., 1986, Liposome electroformation, Faraday Discuss. Chem. Soc., 81.

Arrowsmith, M., Hadgraft, J., and Kellaway, I. W., 1984, The in vivo release of cortisone esters from liposomes and the intramuscular clearance of liposomes, Int. J. Pharm., 20:347.

Bader, H., and Ringsdorf, H., Membrane-spanning symmetric and asymmetric diyne amphiphiles, 1986, Faraday Discuss. Chem. Soc., 81.

Bangham, A. D., Standish, M. M., Watkins, J. C., 1965, Diffusion of univalent ions across the lamellae of swollen lipids, J. Mol. Biol., 13:238.

Barenholz, Y., Amselem, S., and Lichtenberg, D., 1979, A new method for preparation of phospholipid vesicles (liposomes) - french press, FEBS Lett., 99:210.

Batzri, S., and Korn, E. D., 1973, Single bilayer liposomes prepared without sonication, Biochim. Biophys. Acta, 298:1015.

Brunner, J., Skrabal, P., and Hauser H., 1976, Single bilayer vesicles prepared without sonication physico-chemical properties, Biochim. Biophys. Acta, 455:322.

Campbell, Ph. J., 1983, Toxicity of some charged lipids used in liposome preparation, Cytobiol., 37:21.

Cevc, G., Seddon, J. M., and Hartung, R., 1986, Fusogenic liposomes, in: "Liposomes as Drug Carriers", K.-H. Schmidt, ed., Georg Thieme Verlag, Stuttgart - New York.

Chapman, D., Lee, D. C., and Hayward, J. A., 1986, Physicochemical studies of vesicles and biomembranes, Faraday Discuss. Chem. Soc., 81.

Coune, A., Scullier, J. P., Fruhling, J, et al. , 1983, Iv administration of a water-insoluble antimitotic compound entrapped in liposomes. Preliminary report on infusion of large volumes of liposomes to man, Cancer Treat. Rep., 67:1031.

Coune, A., 1984, Lipids and liposomes for improving efficacy of cancer chemotherapy, Eur. J. Cancer, 20:443.

Crommelin, D. J. A., and Van Bommel E. M. G., 1984, Stability of liposomes on storage, Pharm. Res., 4:159.

Cullis, P. R., Hope, M. J., de Kruijff, B., Verkleij, A. J., and Tilcock, C. P. S., 1985, Stuctural properties and functional roles of phospholipids in biological membranes, in: "Phospholipids and Cellular Regulation", Vol. 1, J. F. Kuo, ed., CRC Press, Boca Raton, Florida.

Deamer, D., and Bangham, A.D., 1976, Large volume liposomes by an ether vaporization method, Biochim. Biophys. Acta, 443:629.

Eibl, H., 1981, Phospholipid synthesis, in: "Liposomes: From Physical Structure to Therapeutic Applications", C. G. Knight ed., Elsevier / North-Holland Biomedical Press, Amsterdam - New York - Oxford.

Fidler I. J., and Poste, G., 1982, Macrophage-mediated destruction of malignant tumor cells and new strategies for the therapy of metastatic disease, Springer Semin. Immunpathol., 5:161.

Fromherz, P., Röcker, C., and Rüppel D., 1986, From discoid micelles to spherical vesicles, the concept of edge-activity, Faraday Discuss. Chem. Soc., 81.

Gabev, E. E., Svilenov, D. K., Poljakova-Krusteva, O. T., and Vassilev, J., 1985, Brain, liver and spleen detection of liposomes after rectal administration, J. Microencapsulation, 2:85.

Gerritsen, W. J., Verkley, A. J., Zwaal, R. F. A., and Van Deenen, L. L. M., 1978, Freeze-fracture appearance and disposition of band 3 protein from the human erythrocyte membrane in lipid vesicles, Eur. J. Biochem., 85:255.

Gregoriadis, G. ed., 1984, "Liposome Technology", Vol. III, chapters 2-9, CRC Press, Bota Raton, Florida.

Hayward, J. A., Levine, D. M., Neufeld, L., Simon, S. R., Johnston, D. S., and Chapman, D., 1985, Polymerized liposomes as stable oxygen-carriers, FEBS Lett., 187:261.

Hirnle, P., Jaroni, H., Schubert, R., and Schmidt, K.-H., 1986, Endolymphatic application of liposomal cytostatics for treatment of lymph node metastases: First animal experiments, in: "Liposomes as Drug Carriers", K.-H. Schmidt, ed., Georg Thieme Verlag, Stuttgart - New York.

Hope, M. J., Bally, M. B., Webb, G., and Cullis, P. R., 1985, Production of large unilamellar vesicles by a rapid extrusion procedure. Characterization of size distribution, trapped volume and ability to maintain a membrane potential, Biochim. Biophys. Acta, 812:55.

Huang, C., 1969, Studies on phosphatidylcholine vesicles. Formation and physical characteristics, Biochemistry, 8:344.

Huang, C., and Mason J. T., 1978, Geometric packing constraints in egg phosphatidylcholine vesicles, Proc. Natl. Acad. Sci. USA, 75:308.

Huang, L., 1983, Liposome-cell interactions in vitro, in: "Liposomes", M. J. Ostro, ed., Marcel Deccer, New York - Basel.

Hunt, C. A., and Burnette, R. B., 1984, Lipid microencapsulation of hemoglobin, Appl. Biochem. Biotechnol., 10:147.

Israelachvili, J. N., Marcelja, S., and Horn, R. G., 1980, Physical principles of membrane organization, Q. Rev. Biophys., 13:121.

Killian, J. A., de Kruijff, B., van Echteld, C. J. A., Verkleij, A. J., Leunissen-Bijfelt, J., de Gier, J., 1983, Mixtures of gramicidin and lysophosphatidylcholine form lamellar structures, Biochim. Biophys. Acta, 728:141.

Knight, C.G., Bard, D.R., and Thomas, D. P. P., 1985, Liposomes as carrier of antiarthritic agents, Ann. N. Y. Acad. Sci., 446:415.

Lee, V. H. L., Urrea, P.T., Smith, R. E. and Schanzlin, D. J., 1983, Ocular drug bioavailability from topically applied liposomes, Surv. Ophthalmol., 29:335.

Lelkes, P. I., 1984, Methodological aspects dealing with stability measurements of liposomes in vitro using the carboxyfluorescein-assay, in: "Liposome Technology", Vol. III, G. Gregoriadis, ed., CRC Press, Boca Rato, Florida.

Madden, T. D., and Cullis P. R., 1982, Stabilization of bilayer structure for unsaturated phosphatidylethanolamines by detergents, Biochim. Biophys. Acta, 684:149.

Mayhew, E., and Papahadjopoulos, D., 1983, Therapeutic applications of liposomes, in: "Liposomes", M. J. Ostro, ed., Marcel Deccer, New York - Basel.

Mayhew, E., Lazo, R., Vail, W. J., King, J., and Green A. M., 1984, Characterization of liposomes prepared using a microemulsifier, Biochim. Biophys. Acta, 775:169.

Mazer, N. A., Kwasnick, R. F., Carey, M. C., Benedek, G. B., 1976, Quasielastic light scattering spectroscopic studies of aqueous bile salt, bile salt-lecithin and bile salt-lecithin-cholesterol solutions, Mizellization, Solubilization, Microemulsions, 1:383.

McEvoy, L., Williamson, P., and Schlegel, R. A., 1986, Membrane phospholipid asymmetry as a determinant of erythrocyte recognition by macrophages, Proc. Natl. Acad. Sci. USA, 83:3311.

Milsmann, M. H. W., Schwendener, R. A., and Weder, H.-G., 1978, The preparation of large single bilayer liposomes by a fast and controlled dialysis, Biochim. Biophys. Acta, 512:147.

Müller, K., 1981, Structural dimorphism of bile salt/lecithin mixed micelles, Biochemistry, 20:404.

New York Acad. Sci., 1977, "The use of liposomes in biology and medicine", Conf. Proc.

Olson, F., Hunt, C. A., Szoka, F. C., Vail, W.J., and Papahadjopoulos, D., 1979, Preparation of liposomes of defined size distribution by extrusion through polycarbonate membranes, Biochim. Biophys. Acta, 557:9.

Ostro, M. J., and Giacomoni, D., 1983, Liposomes as a tool in molecular biology: a comparison to other methodologies, in: "Liposomes", M. J. Ostro, ed., Marcel Deccer, New York - Basel.

Pagano, R. E., Schroit A. J., and Struck, D. K., 1981, Interaction of phospholipid vesicles with mamalian cells in vitro: Studies of mechanisms, in: "Liposomes: From Physical Structure to Therapeutic Applications", C. G. Knight, ed., Elsevier / North-Holland Biomedical Press, Amsterdam - New York - Oxford.

Papahadjopoulos, D., Vail, W. J., Jacobson, K., and Poste, G., 1975, Cochleate lipid cylinders: Formation by fusion of unilamellar lipid vesicles, Biochim. Biophys. Acta, 394:483.

Patel, H. M., and Ryman, B. E., 1981, Systemic and oral administration of liposomes, in: "Liposomes: From Physical Structure to Therapeutic Applications", C. G. Knight, ed., Elsevier / North-Holland Biomedical Press, Amsterdam - New York - Oxford.

Pirkl, V., Jaroni, HW., Schubert, R., and Schmidt, K.-H., 1986, Liposome - encapsulated hemoglobin as oxygen carrying blood substitute, J. Eur. Soc. Artif. Org., 4, Suppl. 2:408.

Regen, S. L., 1985, Polymerized phosphatidylcholine vesicles as drug carriers, Ann. N. Y. Acad. Sci., 446:296.

Rhoden, V., and Goldin, S. M., 1979, Formation of unilamellar lipid vesicles of controllable dimensions by detergent dialysis, Biochemistry, 18:4173.

Ring, K., Henkel, B., Valenteijn, A., and Gutermann R., 1986, Studies on the permeability and stability of liposomes derived from a membrane-spanning bipolar archeabacterial tetraetherlipid, in: "Liposomes as Drug Carriers", K.-H. Schmidt, ed., Georg Thieme Verlag, Stuttgart - New York.

Roth, H. J., 1986, Chirale elements of liposomes, in: "Liposomes as Drug Carriers", K.-H. Schmidt, ed., Georg Thieme Verlag, Stuttgart - New York.

Sackmann, E., Duwe, H.-P., and Engelhardt, H., 1986, Membrane bending elasticity and its role for shape fluctuations and shape transformations of cells and vesicles, Faraday Discuss. Chem. Soc., 81.

Scherphof, G., Damen, J., and Hoekstra, D., 1981, Interactions of liposomes with plasma proteins and components of the immune system, in: "Liposomes: From Physical Structure to Therapeutic Applications", C. G. Knight, ed., Elsevier / North-Holland Biomedical Press, Amsterdam - New York - Oxford.

Scherphof, G., van Leeuwen, B., Wilschut, J., and Damen, J., 1983, Exchange of phosphatidylcholine between small unilamellar liposomes and human plasma high-density lipoprotein involves exclusively the phospholipid in the outer monolayer of the liposomal membrane, Biochim. Biophys. Acta, 732:595.

Scherphof, G. L., Damen, J., and Wilschut, J., 1984, Interactions of liposomes with plasma proteins, in: "Liposome Technology", Vol. III, G. Gregoriadis, ed., CRC Press, Boca Raton, Florida.

Schubert, R., Beyer, K., Wolburg, H., and Schmidt, K.-H., 1986a, Structural changes in membranes of large unilamellar vesicles after binding of sodium cholate, Biochemistry, 25:5263.

Schubert, R., Beyer, K., Wolburg, H., Jaroni, H., and Schmidt K.-H., 1986b, Membrane alteration induced by bile salts. Transient membrane holes as one possible mechanism of drug release from liposomes, in: "Liposomes as Drug Carriers", K.-H. Schmidt, ed., Georg Thieme Verlag, Stuttgart - New York.

Schurtenberger, P., Mazer, N., Waldvogel S., and Känzig, W., 1984, Preparation of monodisperse vesicles with variable size by dilution of mixed micellar solutions of bile salt and phosphatidylcholine, Biochim. Biophys. Acta, 775:111.

Schwendener, R. A., Supersaxo, A., Rubas, W., and Weder, H. G., 1986, Liposomes as carriers for lipophilic antitumor prodrugs, in: "Liposomes as Drug Carriers", K.-H. Schmidt, ed., Georg Thieme Verlag, Stuttgart - New York.

Small, D. M., 1971, The physical chemistry of cholanic acids, in "The Bile Acids," Vol. 1, P. P., Nair, and D. Kritchevsky, eds., Plenum Press, New York - London.

Sömjen, G. J., and Gilat, T., 1985, Contribution of vesicular and micellar carriers to cholesterol transport in human bile, J. Lipid Res., 26:699.

Strauss, G., 1984, Freezing and thawing of liposome suspensions, in: "Liposome Technology", Vol. 1, G. Gregoriadis, ed., CRC Press, Boca Raton, Florida.

Stricker, H., and Kibat, P. G., 1986, Storage stability of aqueous liposome dispersions, in: "Liposomes as Drug Carriers", K.-H. Schmidt, ed., Georg Thieme Verlag, Stuttgart - New York.

Sullivan, S. M., Connor, J., and Huang, L., 1986, Immunoliposomes: preparation, properties, and applications, Med. Res. Rev., 6:171.

Szoka, F., and Papahadjopoulos, D., 1978, Procedure for preparation of liposomes with large internal aqueous space and high capture by reverse-phase evaporation, Proc. Natl. Acad. Sci. USA, 75:419.

Träuble, H., and Grell, E., 1971, The formation of asymmetrical spherical lecithin vesicles, Neurosci. Res. Prog. Bull., 9:373.

Weinstein, J. N., and Lesermann, L. D., 1984, Liposomes as drug carriers in cancer chemotherapy, Pharmacol. Ther., 24:207.

Wendel, A., Jeschke, H., and Gloger, M., 1982, Drug-induced lipid peroxidation in mice. II. Protection against paracetamol-induced liver necrosis by intravenously entrapped glutathione, Biochem. Pharmacol., 31:3601.

Westerhof, W., 1985, Possibilities of liposomes as dynamic dosage forms in dermatology, Med. Hypotheses, 16:283.

Wirtz, K. W. A., 1974, Transfer of phospholipids between membranes, Biochim. Biophys. Acta, 344:95.

ABSORPTION AND DISTRIBUTION OF PHOSPHOLIPIDS

Björn Åkesson and Åke Nilsson

Department of Clinical Chemistry and Department
of Medicine, University Hospital, University of
Lund, Lund, Sweden

INTRODUCTION

Phospholipids are ubiquitous components in tissues and in a variety
of foods. The amount and type of phospholipids in the diet may influence
metabolic and physiologic processes in a number of ways, e.g. by influenc-
ing the metabolism and transport of other lipids or by affecting the
formation of different cellular messengers. The absorption and distribution of phosphatidylcholine, the major phospholipid, has been studied
extensively, but less is known about the transport of other phospholipids. In the following recent data on the dietary intake of phospholipids, the transport and metabolism of chyle phospholipids and on the
assessment of phospholipid absorption in man will be briefly summarized.

DIETARY INTAKE OF PHOSPHOLIPIDS

Phosphatidylcholine occurs in many foodstuffs, and the intake of
this substance may be calculated from data on food intake. Wurtman (1979)
computed an intake of 3.1 g phosphatidylcholine from a typical diet in
the United States. In a series of studies on dietary assessment using the
duplicate portion sampling technique, we measured the intake of phospholipids by healthy subjects in Sweden (Åkesson, 1982). Among eight women
the intake of total lipid phosphorus was 1.5-2.5 mmol/day. Phosphatidylcholine (and any ether analogues) constituted 48-70 mole% of the phospholipid intake and the phosphatidylethanolamine fraction was 17-24 mole%.

Vegans have a drastically different intake of dietary lipids as
documented by the duplicate portion sampling technique (Abdulla et al.,
1981). The intake of phospholipids by vegans was 1.5-0.3 mmol/day
(Åkesson, unpublished data). The content of the two major phospholipid
classes was different from that in the common Swedish diet, and phosphatidylcholine was 18-36 mole% and phosphatidylethanolamine 33-47 mole% of
total phospholipids in the vegan diet (Table 1). The lower intake of
phosphatidylcholine among vegans may reflect the absence of eggs, meat
etc in the diet. The significance of this difference is not clear,
although data from animal studies indicate that dietary phospholipids can
exert different physiological effects (Imaizumi et al, 1983).

Dietary phospholipids contain larger proportions of highly unsaturated fatty acids than most other food fats. From data on the usual Swedish

Table 1. CONSUMPTION OF PHOSPHOLIPIDS IN TYPICAL SWEDISH DIETS AND VEGAN DIETS

PHOSPHO-LIPID	TYPICAL DIET	VEGAN DIET
	% OF LIPID P	
PtdEtn	20.7	37.9
PtdCho	62.0	29.3
	MMOL/DAY	
TOTAL PL	2.0	1.5

diet, it could be calculated that the mean consumption of arachidonic acid from phosphatidylethanolamine was 0.09 mmol/day and from phosphatidylcholine 0.08 mmol/day. The intake of docosahexaenoic acid from phosphatidylethanolamine was 0.07 mmol/day and from phosphatidylcholine 0.13 mmol/day (Åkesson, 1982). These fatty acids may also be consumed as triacylglycerols from marine food fats, but in total lipid extracts of typical Swedish diets, the two fatty acids occurred only in small proportions, and no quantitation was made (Borgström et al., 1979). At present it is therefore difficult to evaluate the quantitative importance of dietary phospholipids for the provision of highly unsaturated fatty acids.

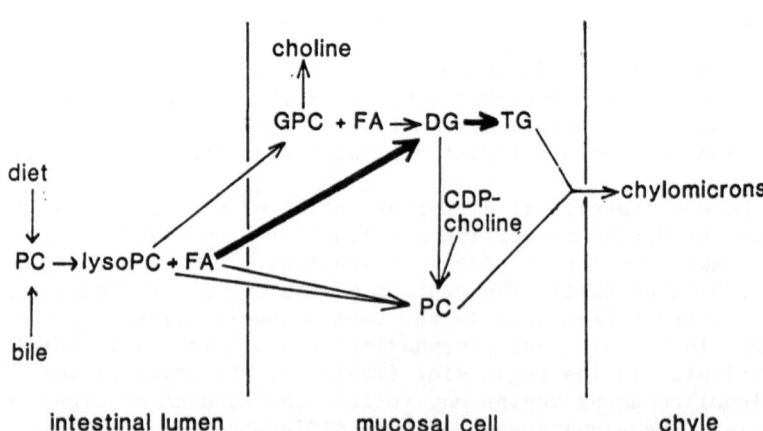

Figure 1. Pathways in the absorption of phosphatidylcholine (PC). GPC, glycerophosphocholine; FA, fatty acid; DG, diacylglycerol; TG, triacylglycerol; CDP-choline, cytidinediphosphocholine.

Table 2. SOURCES OF CHYLE PHOSPHOLIPIDS

1. REACYLATION OF ABSORBED LYSO-PtdCho
2. DE NOVO SYNTHESIS IN THE MUCOSA
3. PROVISION FROM MUCOSAL POOLS
4. ABSORPTION OF INTACT PtdCho

DIGESTION AND ABSORPTION OF PHOSPHOLIPIDS

Dietary phospholipids are mixed with biliary phospholipids, mainly phosphatidylcholine, in the intestinal lumen, and usually the phospholipids originating from bile are in several-fold excess. The metabolic fate of dietary and biliary phosphatidylcholine is probably similar. Most or all of the luminal phosphatidylcholine is cleaved to fatty acid and lysophosphatidylcholine by phospholipase A_2 from the pancreatic juice (Figure 1). The two breakdown products are absorbed across the intestinal mucosa (Scow, Stein and Stein, 1967; Nilsson, 1968), and the mechanisms for this transport will not be dealt with in the present paper.

Later studies challenged this pathway of absorption, claiming that there is an absorption, and indeed an enterohepatic circulation of intact biliary phosphatidylcholine (Boucrot, 1972). Under certain conditions biliary phosphatidylcholine was found to exist as a large complex with biliary protein, rather than as mixed bile salt-phosphatidylcholine micelles only. The further testing of this interesting possibility has, however, not confirmed the presence of such a complex, in which phosphatidylcholine would undergo an enterohepatic circulation. For instance, when rat bile, labeled by injection of (^3H)palmitic acid, was reinfused intraduodenally in other bile fistula rats, no labeled phosphatidylcholine appeared in bile soon after infusion (Larsson and Nilsson, 1978). Zierenberg and Grundy (1982) assessed the intestinal absorption in man of 1 g of $^3H/^{14}C$-labeled phosphatidylcholine and found that more than 90 % was absorbed from the intestine. Although absorption of intact phosphatidylcholine could not be excluded, the most straightforward interpretation of their data is that absorption of lysophosphatidylcholine and fatty acid is the major mechanism.

TRANSPORT OF PHOSPHOLIPIDS IN THE CHYLE

Chyle phosphatidylcholine can be formed by reacylation of absorbed lysophosphatidylcholine or by de novo synthesis from diacylglycerol and CDP-choline, a reaction which increases after fat feeding (Table 2). In addition, some chyle phosphatidylcholine is derived from a preexisting pool of mucosal phospholipids (Arvidson and Nilsson, 1972; Patton et al., 1984). The relative importance of the pathways is influenced by the amount of absorbed lysophosphatidylcholine, originating from both bile and dietary phosphatidylcholine (Mansbach, 1977). The supply of phospholipids to the intestine seems to be necessary for an optimal intestinal lipoprotein secretion, since mucosal cells from rats with biliary drainage tend to accumulate triacylglycerol, and this can be reversed by adding phospholipid (O'Doherty, Kakis and Kuksis, 1973; Tso, Lam and Simmonds, 1978). Different molecular species of phosphatidylcholine are also selectively incorporated into different lipoprotein classes (Patton et al., 1984).

In chylomicrons the proportion of phosphatidylethanolamine is higher than in plasma lipids (Minari and Zilversmit, 1963). Intestinal chylomicrons and VLDL contain apolipoprotein B, but also peptides conforming to an HDL pattern with apo AI as a predominant peptide component. The chylomicron surface material of phospholipids and peptides does not have the same composition as HDL, but the ratio of phospholipid to apo AI is several-fold higher than in circulating HDL. Also HDL produced in the intestine has this high phospholipid/apo AI ratio (Johansson and Nilsson, 1981; Green and Glickman, 1981 and references cited therein). The importance of the amount of dietary phosphatidylcholine for the postprandial composition of plasma lipoproteins in humans was studied by Simonsson, Nilsson and Åkesson (1982). In healthy subjects with intact biliary production of phospholipids the ingestion of a test meal containing 8 g of phosphatidylcholine did not change the phospholipid content of plasma HDL, LDL and VLDL plus chylomicrons up to 6h after a fat meal.

METABOLISM OF CHYLE PHOSPHOLIPIDS

Because of the similarities in the composition of chylomicron surface components and HDL, a role of chylomicrons as donors of HDL material has emerged (Redgrave and Small, 1979). Apo AI is undoubtedly transferred to HDL and metabolized at the same rate as HDL and apo AI from other sources. The intestine may thus be a major contributor of circulating HDL.

In vitro, chylomicron phospholipid is hydrolyzed by the phospholipase A_1 activity of lipoprotein lipase, but in vivo most of the phosphatidylcholine escapes this hydrolysis by a rapid transfer to HDL, where it is resistant to the enzyme. Since chylomicrons contribute relatively more phospholipids than apo AI to plasma HDL, a particulate removal of HDL is unlikely to be the major pathway for phospholipid clearance. Instead an intravascular metabolism of HDL phospholipid, without any simultaneous clearance of apo AI, mediated by the action of LCAT or hepatic lipase is the most likely possibility.

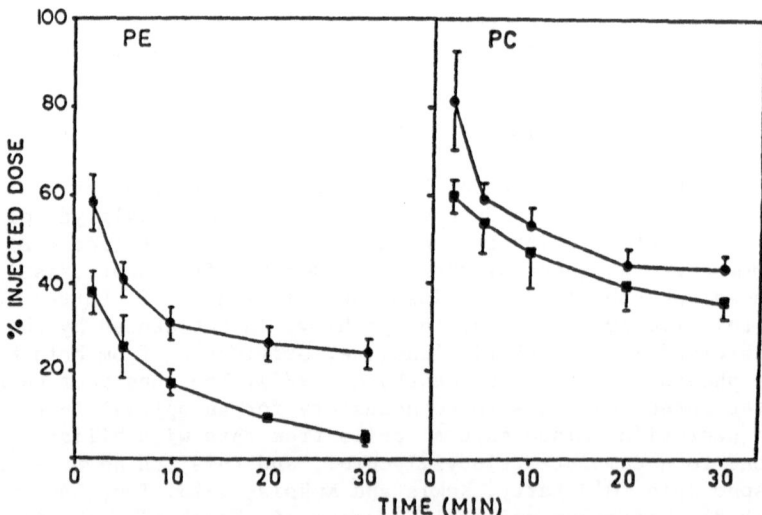

Figure 2. Effect of antiserum against hepatic lipase on the clearance of chylomicron (^{32}P)phosphatidylethanol amine and (^{32}P) phosphatidylcholine. Rats were injected with antiserum against hepatic lipase (●) or control serum (■) 5 min prior to the administration of chylo microns. Values are the means ±SEM for three animals. Reproduced with permission from B. Landin et al.,1984, J Lipid Res., 25:559.

Intestinal lipoproteins provide to the blood a large amount of phospholipid that has to be metabolized. If the action of LCAT was responsible for the clearance of the phospholipid excess of the 30 mg of phosphatidylcholine produced by the intestine per 24 h in the rat, an equimolar amount of HDL-cholesterol ester would be formed. This amount would actually exceed the whole turnover of cholesterol ester in plasma HDL. In addition, arachidonic acid is the predominant fatty acid in rat plasma cholesterol ester, whereas approx. 50 % of the fatty acids at position 2 of chylomicron phosphatidylcholine is linoleic acid. It thus seems that LCAT can only metabolize a limited proportion of chyle phosphatidylcholine, possibly with a preference for species containing arachidonic acid.

An important role for the action of hepatic lipase and lipoprotein lipase in the metabolism of chyle phospholipids seems likely. Injection of anti-serum against hepatic lipase increases the concentration of plasma HDL phospholipids, indicating that the effect of this enzyme on HDL phospholipids in vitro also has a physiological correlate (Jansen, van Tol and Hülsmann, 1980; Kuusi, Kinnunen and Nikkilä, 1979). In recent experiments ^{32}P-labeled chylomicrons were injected intravenously into control rats and into rats treated with anti-serum against hepatic lipase (Landin et al., 1984). A significant effect on the removal of total lipid ^{32}P could not be demonstrated in 30 min, but this time interval may be too short to find significant effects. Chylomicron phosphatidylethanolamine is metabolized faster than total phospholipids, and its clearance is also rapid after its transfer to HDL (Landin and Nilsson, 1984). The clearance of phosphatidylethanolamine after this transfer was significantly inhibited by the anti-serum to hepatic lipase (Figure 2). During the initial phase when phosphatidylethanolamine was still present in the chylomicrons, the anti-serum had no effect on the removal rate, indicating that during this period lipoprotein lipase had an important role (Landin et al., 1984). When chylomicron phosphatidylethanolamine was transferred to HDL in vitro and then injected into rats, the inhibition of the phosphatidylethanolamine clearance was still more marked (Figure 3). In this experiment there was also a 2-2.5-fold increase in the concentration of phosphatidylethanolamine in plasma, indicating that this phospholipid may have a more important role in the transport of polyunsaturated fatty

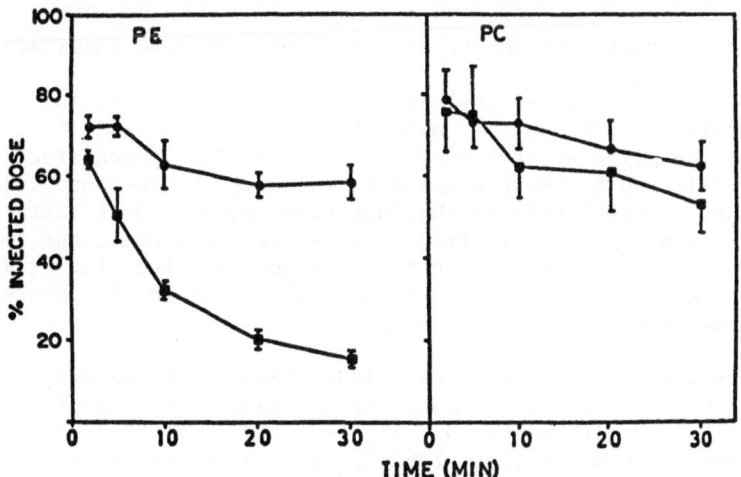

Figure 3. Effect of antiserum against hepatic lipase on the clearance of HDL (^{32}P)phosphatidylethanolamine and (^{32}P)phosphatidylcholine. For details see legend to Figure 2. Reproduced with permission from B. Landin et al., 1984, J Lipid Res, 25:559.

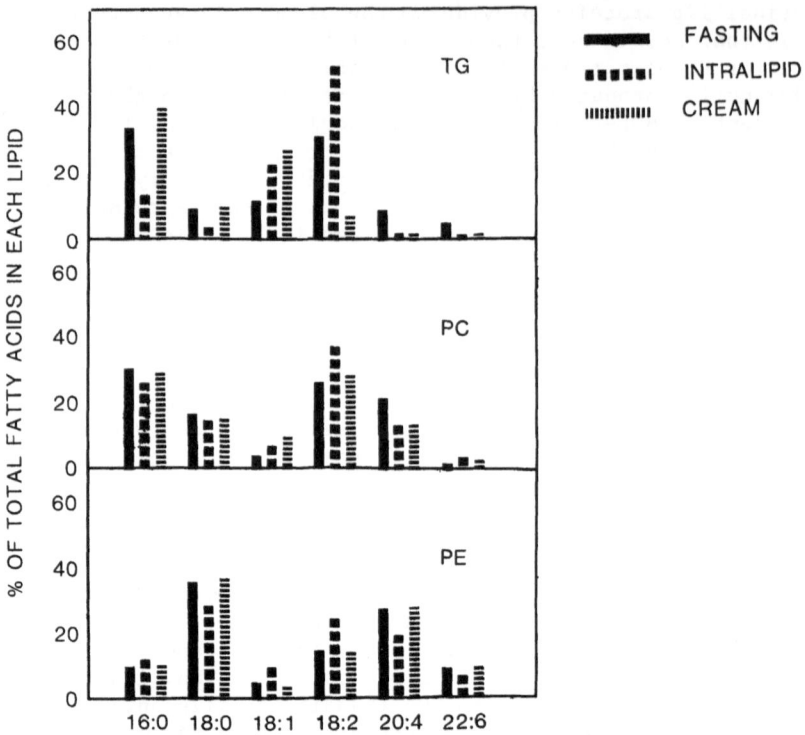

Figure 4. Fatty acid composition of chyle triacylglycerol, phosphatidylcholine and phosphatidylethanolamine in rats in the fasting state and 0-8 h after feeding a soybean oil emulsion (Intralipid[R]) or cream.

acids than is apparent from its low concentration in plasma. Using the rapidly metabolized phosphatidylethanolamine as a marker, the role of hepatic lipase in the clearance of chylomicron phospholipids could thus be clearly demonstrated (See also Landin, 1984, and references cited therein).

ROLE OF CHYLE PHOSPHOLIPIDS IN THE TRANSPORT OF ESSENTIAL FATTY ACIDS

Phospholipids from bile, chylomicrons and other circulating lipoproteins contain large amounts of arachidonic acid, primarily located at position 2 of the glycerol backbone. In the intestinal lumen arachidonic acid will be released by phospholipase A_2 and then absorbed and reincorporated into glycerolipids in the intestinal mucosa. Most fatty acids are primarily reincorporated in triacylglycerol, but despite the important role of arachidonic acid as a precursor of prostaglandins, leukotrienes and other eicosanoids, little attention has been paid to the absorption and lipoprotein transport of this fatty acid.

In recent experiments (Nilsson et al., 1987) these problems were studied in rats with thoracic duct fistulas. (^{3}H)Arachidonic acid and (^{14}C)linoleic acid were mixed with either 0.5 ml cream (40% fat) or 1 ml of a fat emulsion of 20% soybean oil (Intralipid[R]) and the mixtures were fed to rats by stomach tube. More of the given ^{14}C than of ^{3}H was recovered in chyle lipids. At 8-24 h after administration the appearance of ^{3}H in chyle lipids was instead more prominent than that of ^{14}C, leading to an increasing ^{3}H/^{14}C ratio with time. The distribution of radioactivity among different lipid classes varied markedly with time and also with fat vehicle. More of ^{3}H than of ^{14}C was transported with phosphatidylcholine and phosphatidylethanolamine and, in rats fed cream, phospholipids accounted for the major part of ^{3}H in chyle lipid during the absorptive peak at 2-5 h after the meal.

The fatty acid composition of the given fat affected the fatty acid composition of chyle triacylglycerol. After the Intralipid^R meal also the phospholipid fatty acids changed with an increased proportion of linoleic acid (Figure 4). From chemical analysis of the chyle output of lipid classes and their fatty acid composition, the role of different lipids for the transport of arachidonic acid and linoleic acid in chyle was calculated (Figure 5). Phosphatidylcholine was the major lipid transporting arachidonic acid after both types of fat meals.

The mechanism behind the preferential transport of arachidonic acid with phospholipids may be a selective acylation of 1-acyl-phospholipids by arachidonoyl-CoA in the intestinal mucosa. The selectivity was more pronounced, when arachidonic was fed in the fat meal essentially devoid of polyunsaturated fatty acids. It may also be speculated that one reason for the preferential transport of arachidonic acid with phospholipids may be that the further metabolism of chyle phospholipids by lipase hydrolysis at position 1 may provide a system for the transport of arachidonoyl moieties into the tissues without any risk for uncontrolled intravascular release of unesterified arachidonic acid.

ABSORPTION OF PHOSPHATIDYLCHOLINE IN PATIENTS WITH SHORT BOWEL SYNDROME.

In patients with short bowel or ileal disease, a defect in fat absorption may be due to several mechanisms, such as diminished absorptive surface, decrease in bile salt pool size and increased transport rate of intestinal contents. This may lead to changes in the different physical forms of lipids in the intestinal contents and therefore affect differently the absorption of individual lipids, such as triacylglycerol, phospholipids, cholesterol and fat-soluble vitamins. As discussed above, dietary and biliary phospholipids are rich in polyunsaturated fatty acids, and therefore malabsorption of phospholipids may affect the essential fatty

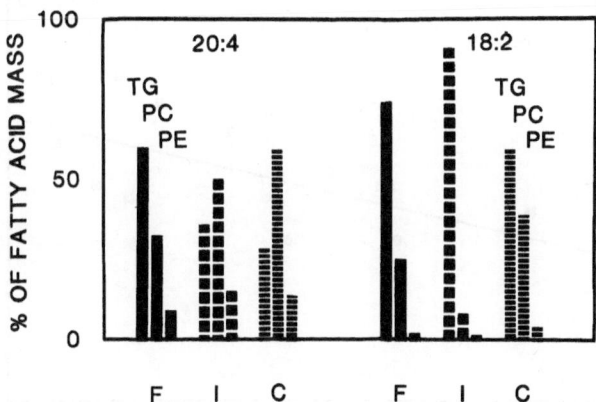

Figure 5. Relative role of chyle triacylglycerol, phosphatidylcholine and phosphatidylethanolamine in the transport of arachidonic acid (20:4) and linoleic acid (18:2) in the fasting state (F) or 0-8 h after feeding a soybean oil emulsion (Intralipid^R, I) or cream (C).

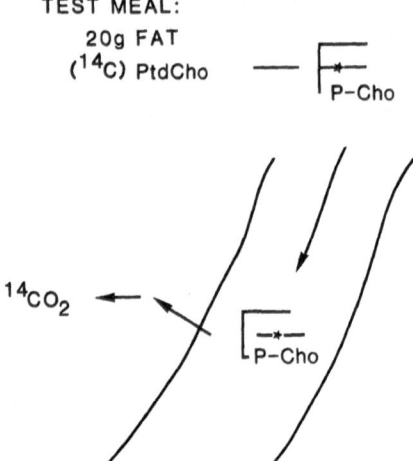

TEST MEAL:
20g FAT
(^{14}C) PtdCho

Figure 6. Outline of the carbon dioxide breath test
for the assessment of the intestinal digestion and
absorption of phosphatidylcholine.

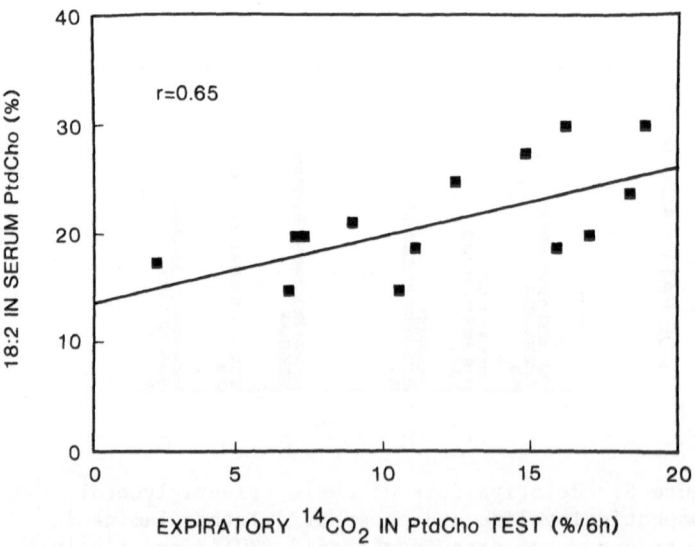

Figure 7. Correlation of the percentage of linoleic
acid in serum phosphatidylcholine and the cumulative
excretion of $^{14}CO_2$ after a (^{14}C)phosphatidylcholine
test meal in patients with short bowel.

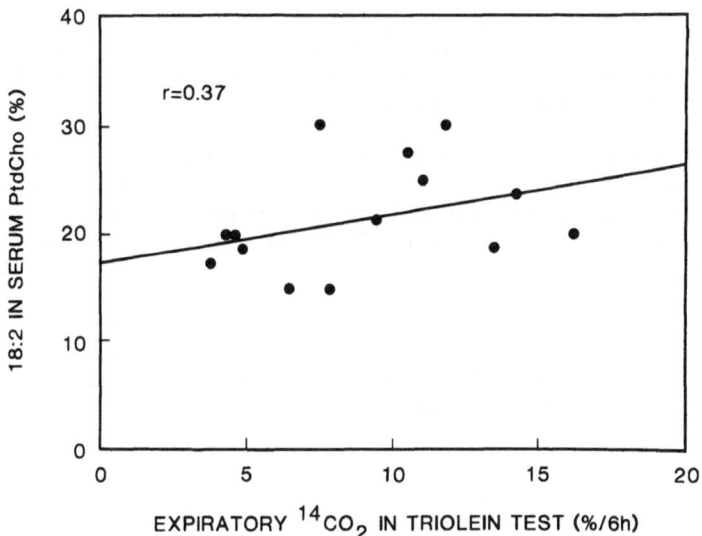

Figure 8. Correlation of the percentage of linoleic
acid in serum phosphatidylcholine and the cumulative
excretion of $^{14}CO_2$ after a (^{14}C)triolein test meal in
patients with short bowel.

acid status in patients. Since very few studies have been performed on
phospholipid absorption in patients, we designed a carbon dioxide breath
test for this purpose (Figure 6) and studied the changes in phospholipid
absorption compared to those in triacylglycerol absorption.

The triolein breath test was performed as described previously
(Åkesson and Florén, 1984). The phosphatidylcholine breath test was made
in an analogous manner except that 65 kBq of 1-palmitoyl, 2-(^{14}C)oleoyl-
sn-glycerol-3-phosphocholine mixed with 100 ml of soybean oil emulsion
(Intralipid^R, 200 mg/ml) was given orally to the subjects. The meal was
given to the subjects after an overnight fast, and breath was sampled at
0, 2, 4, 5 and 6 h (Åkesson and Florén, 1984). For the calculation of
expiratory $^{14}CO_2$ production, a constant endogenous CO_2 production of 9
mmol/kg body weight/h was assumed.

The patients studied had Crohn's disease or had undergone resection
of the small intestine due to Crohn's disease, intestinal volvulus or
occlusion of a mesenterial vein. The controls were healthy subjects
without any known gastrointestinal disease.

In healthy subjects the production of $^{14}CO_2$ from triolein was
somewhat lower than that from (^{14}C)phosphatidylcholine, and the mean
ratio between the data from the two tests increased from 0.7 to 0.9 from
2 to 6 h after the test meal. The absorption of both compounds would be
expected to be close to 100 %, and the difference for the two lipids is
probably explained by differences in absorption kinetics or in the
transport form for the absorbed (^{14}C)oleoyl moiety in chyle lipoproteins,
which may lead to another tissue distribution and percentage conversion
of fatty acid to $^{14}CO_2$. In patients with short bowel the absorption of
(^{14}C)triolein was subnormal, and the cumulative production of $^{14}CO_2$
during 6 h was 9.0±3.9 (n=15) % of administered dose (mean±SD) compared
to 14.4±3.1 (n=7) among the control subjects. Furthermore the absorption
of (^{14}C)phosphatidylcholine was depressed, and the production of $^{14}CO_2$
was 11.8±4.9 %/6h compared to 19.9±3.7 %/6h among controls. Also among

the patients the production of $^{14}CO_2$ from (^{14}C)phosphatidylcholine was higher than that from (^{14}C)triolein, and there was no evidence for a selective derangement in the absorption either of triacylglycerol or phosphatidylcholine among the patients.

Essential fatty acid status of the patients was assessed by analysis of the fatty acid composition of serum phosphatidylcholine. There was a significant positive correlation between the percentage of linoleic acid and the cumulative $^{14}CO_2$ production after the (^{14}C)phosphatidylcholine test meal (Figure 7). The relation was less apparent with the $^{14}CO_2$ production from (^{14}C)triolein (Figure 8). This indicates that the extent of phosphatidylcholine and triolein malabsorption are important for the maintenance of essential fatty acid balance in this group of patients. It also points to a role for dietary and biliary phospholipids for the provision of essential fatty acids.

CONCLUSIONS

Most of the phospholipids in the body are located in cellular membranes, and the fatty acid composition of each phospholipid is one important parameter for different membrane functions. The data reviewed here, show that the association of a specific fatty acid composition with phospholipids may be important also for the provision of polyenoic fatty acids with the diet, and for the transport of different essential fatty acids in circulating lipoproteins. Of special interest is the finding that a large proportion of exogenous and endogenous arachidonic acid, transported in chyle of rats, is located in phosphatidylcholine and phosphatidylethanolamine. The latter phospholipid may play a greater role for the transport of essential fatty acids than is indicated by its concentration in lipoproteins.

It is important to establish methods for the study of absorption and transport of essential fatty acids and phospholipids also in man. For this purpose we have worked out a phosphatidylcholine breath test for the assessment of phospholipid absorption.

ACKNOWLEDGEMENTS

Dr E. Jensen, Dr B. Landin, Ms B. Mårtenssen, Ms G. Olsson, Ms B. Persson and Ms B. Sundén made important contributions to the studies from the authors' laboratories. Financial support was obtained from the Medical Research Council (projects 3968 and 3969), the Nutrition Foundation of the Swedish Margarine Industry, A. Påhlsson's Foundation and the Medical Research Council of the Swedish Life Insurance Companies.

REFERENCES

Abdulla, M., Andersson, I., Asp, N.G., Berthelsen, K., Birkhed, D., Dencker, I., Johansson, C.G., Jägerstad, M., Kolar, K., Nair, B.M., Nilsson-Ehle, P., Nordén, Å., Rassner, S., Åkesson, B., and Öckerman, P.A., 1981, Nutrient intake and health status of vegans. Chemical analysis of diets using the duplicate portion sampling technique,
Amer. J. Clin. Nutr., 34:2464.
Åkesson, B., 1982, Content of phospholipids in human diets studied by the duplicate portion technique,
Br. J. Nutr., 47:223.

Åkesson, B., and Florén, C.H., 1984, Use of the triolein breath test for the demonstration of fat malabsorption in coeliac disease, Scand. J. Gastroenterol., 19:307.

Arvidson, G. A. E., and Nilsson, Å., 1972, Formation of lymph chylomicron phosphatidylcholines in the rat during absorption of safflower oil or triolein, Lipids, 7:344.

Borgström, B., Nordén, Å., Åkesson, B., Abdulla, M., and Jägerstad, M., 1979, Nutrition and old age. Chemical analyses of what old people eat and their states of health during 6 years of follow-up, Scand. J. Gastroenterol., 14, suppl. 52:1.

Boucrot, P., 1972, Is there an entero-hepatic circulation of the bile phospholipids?, Lipids, 7:282.

Green, P. M. R., and Glickman, R. M., 1981, Intestinal lipoprotein formation, J. Lipid. Res., 22:1153.

Imaizumi, K., Mawatari, K., Murata, M., Ikeda, I., and Sugano, M., 1983, The contrasting effect of dietary phosphatidylethanolamine and phosphatidylcholine on serum lipoproteins and liver lipids in rats, J. Nutr., 113:2403.

Jansen, H., van Tol, A., and Hülsmann, W.C., 1980, On the metabolic function of heparin releasable liver lipase, Biochem. Biophys. Res. Commun., 92:53

Johansson, B., and Nilsson, Å., 1981, Plasma apolipoprotein A-I levels after thoracic duct drainage in the rat, FEBS Lett., 130:305.

Kuusi, T., Kinnunen, P. K. J., and Nikkilä, E.A., 1979, Hepatic endo-thelial lipase antiserum influences rat plasma low and high density lipoprotein in vivo, FEBS Lett., 104:384.

Landin, B., 1984, Studies on the hepatic metabolism of chylomicron phospholipids and chylomicron remnants in the rat, Thesis, University of Lund.

Landin, B., Nilsson, Å., Twu, J. S., and Schotz, M.C., 1984, A role for hepatic lipase in chylomicron and high density lipoprotein phospholipid metabolism, J. Lipid. Res., 25:559.

Landin, B., and Nilsson, Å., 1984, Metabolism of chylomicron phosphatidylethanolamine in the rat, Biochim. Biophys. Acta, 793:105.

Larsson, B., and Nilsson, Å., 1978, Lack of enterohepatic circulation of intact biliary phosphatidylcholine in the rat, Scand. J. Gastroent, 13:273.

Mansbach, C.M.,1977, The origin of chylomicron phosphatidylcholine in the rat, J. Clin. Invest., 60:411.

Minari, O., and Zilversmit, D.B., 1963, Behavior of dog lymph chylomicron lipid constituents during incubation with serum, J. Lipid. Res., 4:424.

Nilsson, Å., 1968, Intestinal absorption of lecithin and lysolecithin by lymph fistula rats, Biochim. Biophys. Acta, 152:379.

Nilsson, Å., Landin, B., Jensen, E., and Åkesson, B., 1987, Absorption and lymphatic transport of exogenous and endogenous arachidonic and linoleic acid in the rat, Amer. J. Physiol., 252:

O'Doherty, P.J.A., Kakis, G., and Kuksis, A., 1973, Role of luminal lecithin in intestinal fat absorption, Lipids, 8:249.

Patton, G.M., Bennet Clark, S., Fasulo, J. M., and Robins, S. J., 1984, Utilization of individual lecithins in intestinal lipoprotein formation in the rat, J. Clin. Invest., 73:231.

Redgrave, T. G., and Small, D. M., 1979. Quantitation of the transfer of surface phospholipid of chylomicrons to the high density lipoprotein fraction during the catabolism of chylomicrons in the rat, J. Clin. Invest., 64:162.

Scow, R.O., Stein, Y., and Stein, O., 1967, Incorporation of dietary lecithin and lysolecithin into lymph chylomicrons in the rat, J. Biol. Chem., 242:4919.

Simonsson, P., Nilsson, Å., and Åkesson, B., 1982, Postprandial effects of dietary phosphatidylcholine on plasma lipoproteins in man, Amer. J. Clin. Nutr., 35:36.

Tso, P., Lam, J., and Simmonds, W. J., 1978, The importance of the lysophosphatidylcholine and choline moiety of bile phosphatidylcholine in lymphatic transport of fat, Biochim. Biophys. Acta, 528:364.

Wurtman, J. J., 1979, Sources of Choline and Lecithin in the Diet, in: "Nutrition and the Brain", vol. 5, A. Barbeau, J. H. Growdon, R. J. Wurtman, ed., pp. 73-81, Raven Press.

Zierenberg, O., & Grundy, S. M., 1982, Intestinal absorption of polyenephosphatidylcholine in man, J. Lipid. Res., 23:1136.

CONTROL OF LECITHIN METABOLISM

Dennis E. Vance

Lipid and Lipoprotein Research Group and
Department of Biochemistry
Faculty of Medicine, University of Alberta
Edmonton, Alberta T6G 2C2

INTRODUCTION

The discovery of lecithin has been attributed to M. Gobley in 1847. Lecithin derives it's name from the Greek word *lekithos* that means egg yolk. Remarkably, the stoichiometry of the chemical composition of lecithin was elucidated in the 1860's by Diacknow and Strecker and the chemical synthesis of lecithin was achieved in 1927 by Grun and Limpacher. As has been the case for research in other areas of biology, once the chemistry of a natural product was elucidated, the stage was set for elucidating its biosynthesis. I usually attribute the biochemistry to have begun with the publication in 1932 by Best and Huntsman that choline in the diet would alleviate fatty livers that resulted from lecithin deficiency[1]. The next major development was Stetten and du Vigneaud's demonstration in 1941 that it was possible to convert phosphatidylethanolamine (PE) to phosphatidylcholine (PC) in rat liver.

The enzymology of PC metabolism began when the DNA enzymologist Arthur Kornberg and J. Wittenburg described choline kinase activity in yeast, liver and other mammalian tissues[2]. Subsequently Eugene Kennedy and co-workers described the reactions catalyzed by CTP:phosphocholine cytidylyltransferase (CT) and CDP-choline: 1, 2-diacylglycerol cholinephosphotransferase[3]. Thus, by 1956, the enzymatic pathway for the conversion of choline to PC was understood. In 1961 the microsomal enzymatic activity responsible for the conversion of PE to PC was described by Bremer and Greenberg[4]. Since that time much effort has been expended to purify the four enzymes involved in PC biosynthesis. It is only in the last two years that purification to homogeniety has been reported for three of these enzymes, as will be metnioned below. Although the cholinephosphotransferase has been solubilized[5], it remains an intractable enzyme for purification to homogeniety.

The major objective of this review is to describe the current status of our knowledge about the regulation of lecithin synthesis via the CDP-choline pathway. However, the first selection will describe the status of the proposal that methylation of PE has a physiological role in signal transmission[6].

LECITHIN SYNTHESIS VIA THE METHYLATION OF PHOSPHATIDYLETHANOLAMINE

Enzyme Characterization

The major problem in understanding the enzymology of the conversion of PE to PC has been the purification of the microsomal enzyme responsible for this transformation. In 1979 a partial purification was reported in which the enzyme was solubilized from microsomes by Triton X-100[7]. Unfortunately the enzyme activity was unstable. More recently Mato's laboratory has reported an improved purification of the enzyme from rat liver[8]. Their solubilization involved a buffer that contained 0.3% CHAPS and anion exchange chromatography followed by gel filtration. The apparent success of their preparation involved an incubation with LiBr which dissociated the activity from the lipid. Although there was a single band on SDS electrophoresis with a molecular weight of about 50,000, the specific activity was only 273 nmoles of methyl group transferred per min per mg of protein. A specific activity between 10 to 100 times higher would be typical for a purified phospholipid synthetic enzyme.

Neale Ridgway working in my laboratory has recently achieved purification of PE methyltransferase to homogeneity (submitted for publication). The enzyme has a single subunit with a molecular weight of 18,300. The enzyme catalyzes all three transmethylation reactions in the conversion of PE to PC. The specific activities of the enzyme with PE, monomethyl-PE and dimethyl-PE as substrates were 0.63, 8.59 and 3.75 μmol/min/mg protein, respectively. Thus, 26 years after its discovery[4], this enzyme has finally been purified to homogeneity. The door is now open for numerous structural and regulatory studies on this enzyme.

Physiological Function of PE Methylation

The methylation of PE is 10-to 1000-fold higher in liver than reported in other tissues of the rat[9]. Secondly, it has been estimated that in rat hepatocytes, the methylation pathway accounts for ~ 20% of PC made in that organ[10]. It, therefore, seems that a clear function of PE methylation in rat liver is to make PC.

Even though the activity in other cells and tissues is comparatively low, the evidence is convincing that a trace amount of activity is indeed present. This was most vivdly demonstrated by the 1548-fold purification from mouse thymus of the methylation activity[11]. The function, if any, of this activity is not presently known. Perhaps it is simply present because the gene coding for this activity is not totally repressed and the trace activity has no significant purpose for the cell.

The earlier proposal by Hirata and Axelrod[6] that PE methylation was an integral part of biological signal transmission generated much interest and controversy. Their hypothesis would have provided an excellent explanation for the low activity present in many cells. Although a stimulating proposal, many laboratories have been unable to show a linkage between the transmission of biological signals at the cell surface and PE methylation[12].

Thus, the current evidence suggests that the only function of PE methylation in mammalian systems is to make PC in the liver.

Enzymology of the CDP-Choline Pathway

The initial reaction in PC biosynthesis is catalyzed by choline kinase. In this reaction choline is phosphorylated by Mg^{+2}-ATP to yield cholinephosphate and ADP. This reaction commits choline to the biosynthesis of PC in all animal tissues.

Since the discovery of this enzyme from yeast, many attempts have been made for the purification of this enzyme from cytosol. Early efforts yielded 700-fold purified preparations from rat liver using affinity chromatography[13]; but homogeneity was not obtained. However, when Ishidate and coworkers turned to kidney, they were successful in obtaining a homogenous enzyme[14]. The enzyme appears to be a dimer of subunits with a molecular weight of 42,000. Contrary to previous beliefs the purified enzyme also contained ethanolamine kinase activity[15], strongly indicating that both kinase activities reside on the same protein. Antibodies were raised against the rat kidney enzyme and almost completely inhibited both the choline kinase and ethanolamine kinase activity in cytosols recovered from lung, intestine and liver[15]. However, kinetic data suggest that the active sties for the two kinase activities are separate.

Two other points about choline kinase should be mentioned. First, the enzyme in liver and other tissues has been found in three different forms which differed in apparent molecular size and electrophoretic mobility on native or isoelectric focusing polyacrylamide gels[16]. The physiological significance of these three forms needs to be elucidated. Second, choline kinase can be induced in liver with excess choline, insulin, diethylstilbesterol, polycyclic aromatic hydrocarbons and carbon tetrachloride. The latter two treatments result in the synthesis of a form of the enzyme with is normally prsent only at low levels in the livers of untreated rats[16]. Choline kinase seems to be the only enzyme in the CDP-choline pathway that is regulated at the level of gene expression.

The second enzyme in this biosynthetic sequence is CTP:phosphocholine cytidylyltransferase (CT). This enzyme has the rather unusual, but not unique, property of being located both in the cytosol and microsomes. Early efforts at purification of the cytosolic activity were frustrated by aggregation of the enzyme and loss of activity after DEAE chromatography. In 1977 Choy, Lim and Vance reported a purification of the enzyme to a single band on non-denaturing gels[17]. At that time many of the properties of the enzyme were described. The most unusual property of the preparation was the absolute requirement of phospholipids for activity. The best activators were lyso-PE, and the anionic phospholipids phosphatidylserine, phosphatidylglycerol and phosphatidylinositol[18]. The activation by lyso-PE could not be attributed to detergent effects since lyso-PC did not activate CT, but inhibited it. Other detergents had little or no effect on the enzyme activity[18].

The importance of CT in regulation of PC biosynthesis was recognized in the late 1970's when it was found that CT catalyzed the rate-determining step for PC biosynthesis in many different types of cells and tissues[19]. The first evidence that this might be the case came out of Akesson's laboratory when he measured the concentration of the intermediates for PC biosynthesis in rat liver [20]. They found that the pool size of phosphocholine in rat liver (~14 umoles/10 g liver) was about 140 times the CDP-choline pool (~0.09 umoles/ 10 g liver). These results are consistent with the metabolic slow step in the sequence being the conversion of phosphocholine to CDP-choline. Pulse-chase studies from Akesson's[10] and our[19] laboratories subsequently demonstrated an accumulation of radioactivity in phosphocholine that was quantitatively converted to PC during the chase period with minimal accumulation of label in CDP-choline. The radioactivity in the total CDP-choline pool was always so low we sometimes wondered if it were an intermediate. However, measurement of the specific radioactivity of CDP-choline showed it approached the same value as for phosphocholine, thus indicating that CDP-choline was most likely an intermediate. Subsequent studies have shown that control of lecithin biosynthesis is centered on the CT reaction as will be discussed in more detail below.

Great efforts have been expended in the past 10 years on the purification to homogeneity of CT. Thus, it was with a sense of relief that people greeted the publication of a purification procedure from Weinhold *et al.* [21]. CT was purified 2180-fold from rat liver cytosol to a specific activity of 12,250 nmoles/min/mg of protein. They reported the enzyme consisted of two subunits with a molecular weight of 39,000 and 48,000. Studies by gel filtration suggest that the purified enzyme was composed of 4 subunits. We have been able to reproduce the purification procedure. Subsequent chromatography of the enzyme on an anion exchange column (mono Q from Pharmacia) removed the low molecular weight subunit without loss of specific activity of the enzyme. Hence, it seems in our hands that only one subunit is required for CT activity. The achievement by Weinhold's laboratory now sets the stage for many new types of studies on CT. Preparation of antibodies to CT will facilitate studies on measurements of the amounts of CT in tissues and studies on phosphorylation/dephosphorylation of CT. In addition antibodies and sequence data on the enzyme will make cloning of the cDNA for CT possible and subsequently the determination of the primary sequence of the enzyme.

The final enzyme in the synthesis of lecithin is the cholinephosphotransferase, an enzyme firmly associated with the endoplasmic reticulum. The enzyme faces the cytosolic surface as suggested by its sensitivity to proteolytic digestion[22]. The enzyme remains refractory to purification despite some heroic attempts in many different laboratories [23]. The acyl specificity of the cholinephosphotransferase has been studied extensively[23]. It appears that the enzyme does not display significant specificity but rather can utilize the entire pool of diacylglycerol species.

The rate of lecithin biosynthesis does not appear to be influenced by the amount or activity of the cholinephosphotransferase. Instead there appears to be an excess of this enzyme in cells. The rate of the cholinephosphotransferase reaction appears to be regulated by the supply of CDP-choline and diacylglycerol [24].

Evidence for the Regulation of the Cytidylyltransferase by Phosphorylation/ Dephosphorylation

The development of a system for the preparation of rat liver hepatocytes in our laboratory set the stage for many different types of experiments on control of PC synthesis in cultures of primary liver cells. Steven Pelech, working in my laboratory, was well aware of the effects of cAMP on other biosynthetic pathways and therefore wondered if this second messenger might not also inhibit PC biosynthesis. Using stable analogues of cAMP and pulse-chase experiments with [^{3}H]choline Steve was able to demonstrate that these compounds did inhibit by about 50% the conversion of phosphocholine to CDP-choline in cultured hepatocytes[25]. Since cAMP activates a specific protein kinase, subsequent studies were directed toward seeing if conditions that favor phosphorylation would inactivate CT and conditions that favor dephosphorylation would activate CT.

In one set of experiments rat liver cytosol was incubated with + or - 0.5mM Mg^{+2} ATP and the activity of CT was followed as a function of time. The control incubations showed an increase in CT activity whereas enzyme activity in those incubations with ATP did not increase during the experiment. The activation of CT in the control incubations reached 5-fold, but could be almost completely blocked by the addition of NaF, an inhibitor of protein phosphate phosphatases[26]. When commercial preparations of protein kinase inhibitors were included in the incubations, the inhibitory effect of ATP was prevented[26]. Finally, when commercial preparations of hog intestinal alkaline phosphatase were added to rat liver cytosol, there was a rapid 4-fold activation of CT activity. These studies complemented the studies in intact hepatocytes with cAMP analogues and together provided compelling, but indirect, evidence that CT is regulated in liver by reversible phosphorylation. The experiments do not demonstrate whether the phosphorylation event is directly on CT or perhaps on another modulator protein. For example, the results would be consistent with the inactivation by phosphorylation of an activator protein that is also in cytosol. Alternatively, there may be a protein phosphate phosphatase inhibitor which might be activated by phosphorylation and thereby prevent the dephosphorylation and hence activation of CT. Nevertheless, it is our working hypothesis that the phosphorylation event is directly on CT. With the availability of pure enzyme, it should be possible to test directly this hypothesis.

It was mentioned above that CT required phospholipids for activity. If this were the case, why is CT activity present in rat liver cytosol in the absence of exogenously added phospholipids? It is because in our preparations of rat liver cytosol, there was always some phospholipid present. It is our hypothesis that dephosphorylation *per se* will not activate CT. Rather, the removal of a phosphate group alters the affinity of the enzyme for the phospholipid present in the cytosol. Once it binds to the lipid, it is activated. This proposal has also not been substantiated yet with purified enzymes and phospholipids.

A Role for Fatty Acids in the Regulation of Lecithin Biosynthesis

It has been recognized for sometime that fatty acids promote the biosynthesis of PC. Sundler and Akesson's now classic study[10] on control of phospholipid synthesis in cultured hepatocytes also had data to indicate that PC synthesis in this system was stimulated by fatty acids. Oleic acid (1 mM) stimulated incorporation of ^{32}P by approximately 3-fold. We became interested in the mechanism by which the fatty acid caused this effect. We confirmed that oleate did indeed stimulate PC synthesis in monolayers of rat hepatocytes[27]. Long chain saturated and unsaturated fatty acids caused the stimulation, but short chain fatty acids such as 8:0 were ineffective. Stimulation of PC synthesis was evident within 30 min after the addition of the fatty acid. The effect was correlated with a doubling of the microsomal CT activity. *In vitro* studies showed that fatty acids also caused the stimulation of CT activity in cytosol. This was attributed to the fatty acid promoting the binding of CT to phospholipids in the cytosol. Fatty acids did not activate CT in delipidated cytosol.

It was at this time that we began to recognize a pattern in the correlation between the rate of PC biosynthesis and the amount of CT bound to microsomal membranes. In the studies with the cAMP analogues, we had observed a decrease in the rate of PC synthesis and a lower specific activity of microsomal CT[25]. The adenosine analogue 3-deaza-adenosine stimulated PC biosynthesis by 2-3 fold, and this correlated with about a 2-fold increase in the specific activity of microsomal CT[28]. Phospholipase C treatment of myoblasts caused an apparent 3-fold increase in PC synthesis with a parallel increase in

CT activity in the microsomes[29]. Other correlations between PC synthesis and CT activity in the rat lung after premature delivery[30] and the stimulation of PC synthesis in rat liver by a cholesterol/cholate diet[31] supported the idea that the active species of CT in cells was the microsomal enzyme and the CT activity in the cytosol was not active, but simply a reservoir. This idea was consistent with our knowledge that CT activity was virtually absent in cytosols from any tissue in which the lipid had been removed.

Nevertheless, a major problem we have faced in our studies on control of PC biosynthesis still made us cautious in drawing any firm conclusions. That problem was that the changes we and others had observed in PC biosynthesis were always very small, in the order of 2 to 3-fold. There was no doubt that we always achieved statistical significance in our experiments. But the dramatic changes that can be seen in fatty acid synthesis, for example the 20-50 fold increase in fatty acid synthesis observed after starvation and refeeding a rat, were never observed in PC metabolism.

One day Steven Pelech's curiosity paid an unexpected dividend. While he was engaged in the studies on the mechanism by which fatty acids activate PC synthesis in hepatocytes, he was conducting other experiments in HeLa cells to understand the mechanism by which phorbol esters activate PC synthesis [32]. He wondered if fatty acids would also stimulate PC synthesis in HeLa cells. The first experiment yielded no effect. Since the fatty acids were bound to albumin when added to the heptocytes, Steven reasoned that albumin may have blocked an effect in HeLa cells. When he added oleate as a suspension to the HeLa cell medium, he observed a dramatic increase in the conversion of phosphocholine to PC (~10-fold)[33]. This was the first time I had ever seen such an increase in PC biosynthesis. The result had important implications for the proposal that CT was activated in cells by translocation of the enzyme from the cytosol to microsomal membranes. It was with great anticipation that Steve treated the HeLa cells with oleate and then fractionated the cells into microsomes and cytosol and assayed for CT activity. The results were unambiguous. Whereas in control cells about 50% of the enzyme is in cytosol and the remainder bound to microsomes, all of the CT activity in oleate treated cells was found on the membranes[33].

The Translocation Hypothesis

The above results led us to formulate the translocation hypothesis[34]. We suggested that CT is found both in the cytosol and microsomes of cells and tissues. The active species is the CT bound to the membranes and the inactive species is in the cytosol. At least two mechanisms exist whereby the cell regulates the amount of CT bound to the membranes. In one case we suggested that the binding to membranes can be regulated by reversible protein phosphorylation. We postulated that a phosphatase exists in the cell which will enhance the enzyme's ability to bind to phospholipid in membranes where it is activated. The inactivation and release of the enzyme can be caused by the action of a cAMP dependent protein kinase. As stated in an earlier section, this proposal needs to be verified with pure CT, pure protein kinase and protein phosphatase.

Secondly, we proposed with considerably more evidence that fatty acids will cause the CT to bind to membranes in the cell and this can be reversed. The current evidence suggests that this might be a mechanism that overrides the inhibitory effect of protein phosphorylation. Although nothing has been published in the last two years that argues against the translocation hypothesis, we still consider it to be a working idea that will be modified as more information becomes available.

Mechanistic Studies on the Translocation of CT by Fatty Acids

In the past several years Rosemary Cornell working in my laboratory has studied in more detail the mechanism by which fatty acids translocate CT in HeLa cells[35, 36]. She confirmed the earlier report by Pelech *et al.* that saturated fatty acids were ineffective in stimulating PC synthesis in HeLa cells but, *in vitro*, palmitic acid was an effective agent for promoting the binding of CT to membranes[33]. We did not understand this apparent contradiction between the studies with intact cells and cell free systems. After the obvious explanations were eliminated, Rosemary noted that there was 2mM calcium in the medium and that might be binding two saturated fatty acids and preventing them from interacting with CT. The addition of EGTA to the medium did indeed result in the stimulation of PC biosynthesis in intact HeLa cells by 1 mM palmitate, a concentration that previously had no effect[35]. Thus, calcium in the medium specifically inhibited the stimulation of PC synthesis by palmitate. The selective effect of calcium on palmitic but not oleic acid is probably due to the ability of saturated but not unsaturat'd fatty acids to pack tightly and form a disoap, Ca^{+2} dipalmitate.

A number of other conclusions resulted from Rosemary's studies. 1. The CT-membrane interaction is hydrophobic. It can be dissociated by detergents, but not by high salt. The association does not seem to require a negative charge since oleoyl alcohol and mono-olein were effective in promoting binding to membranes. 2. The compounds which promote translocation are all amphipathic with one or two long hydrocarbon chains which may be saturated or unsaturated. All the translocators have a free hydroxyl function. 3. The previous report that oleoyl-CoA caused translocation is probably incorrect[27]. Rather it seems that the CoA derivative was being hydrolyzed to oleic acid at a rapid rate and the free acid was responsible for the translocation. 4. The CT-membrane interaction promoted by oleate is relatively specific. Large changes in the distribution of cellular protein between cytosol and membranes do not occcur. 5. The fatty acid-dependent translocation of CT is reversible offering a rapid mechanism for the modulation of PC synthesis in cells.

The nature of the fatty acid mediated binding of CT to membranes was evaluated further in studies with HeLa cell cytosol and large unilamellar vesicles of egg PC or HeLa cell phospholipids[36]. A fatty acid/phospholipid mole ration of 0.1 was required before any binding of CT to the liposomes was detected. At a fatty acid/phospholipid mole ratio of 1, 85% of cytosolic CT was bound to the vesicles. CT binding to the lipid was also promoted by greater than 20 mole percent of palmitic acid, monoolein, diolein or oleoylacetylglycerol in phospholipid vesicles. Oleoyl-CoA did not cause CT to bind to liposomes. Consistent with the studies with cytosol and microsomes, CT binding was blocked with Triton X-100 but not by 1 M KCl. Somewhat surprisingly, CT binding to vesicles was independent of temperature; binding occurred equally well at 4° or at 37°. CT also bound to dimyristoyl- and dipalmitoyl-PC that contained oleic acid at temperatures below the phase transition for these liposomes.

SUMMARY AND CONCLUSIONS

The above experiments and studies from other laboratories have shown many of the requirements for the interactions among CT, membrane phospholipids and fatty acids or related compounds. A number of important questions remain. We still do not understand how the fatty acids actually facilitate the binding of the protein to the lipid vesicles. Knowledge of the sequence of CT should enable us to make predictions about which part of the protein might bind to the fatty acids and to the phospholipids. Another problem relates to the localization of CT in the cell. From our liposome studies it is apparent that the only requirement for CT binding to a membrane is for phospholipid, yet CT is not recovered in all parts of the cell. For example, CT is found on the endoplasmic reticulum and Golgi apparatus, but is not localized to the mitochondria (J. E. Vance and D.E. Vance, unpublished results). Why? Finally, it is clear from studies in our laboratory and other laboratories that the fatty acid mediated translocation of CT can occur in a reversible fashion in cultured cells. How important is this phenomenon in intact animals? One study has addressed this issue[30]. Weinhold *et al.* have reported evidence which correlates increased binding of CT to microsomal membranes in fetal lung *in vivo* with an increase in the fatty acid content of the membranes. Clearly, further evidence of the physiological significance of CT translocation is required.

ACKNOWLEDGEMENT

The research from this laboratory summarized in this article has been supported by research grants from the Medical Research Council of Canada.

REFERENCES

1. C.H. Best and M.E. Huntsman, The Effects of the Components of Lecithine upon Deposition of Fat in the Liver, J. Physiol. London 75: 405-412, (1932).

2. J. Wittenberg and A. Kornberg, Choline Phosphokinase, J. Biol. Chem. 202: 431-444 (1953).

3. E.P. Kennedy, The Metabolism and Function of Complex Lipids, The Harvey Lectures 57: 143-171 (1962).

4. J. Bremer and D.M. Greenberg, Methyltransferring Enzyme System of Microsomes in the Biosynthesis of Lecithin (Phosphatidylcholine), Biochim. Biophys. Acta 46: 205-216 (1961).

5. R. Cornell and D.H. MacLennan, Solubilization and Reconstitution of Cholinephosphotransferase from Sarcoplasmic Reticulum: Stabilization of Solubilized Enzyme by Diacylglycerol and Glycerol, Biochim. Biophys. Acta 821: 97-105 (1985).

6. F. Hirata and J. Axelrod, Phospholipid Methylation and Biological Signal Transmission, Science, 209: 1082-1090 (1980).

7. W.J. Schneider and D.E. Vance, Conversion of Phosphatidylethanolamine to Phosphtidylcholine in Rat Liver, Partial Purification and Characterization of the Enzymatic Activities, J. Biol. Chem. 254: 3886-3891.

8.   M.A. Pajares, M. Villalba and J.M. Mato, Purification of Phospholipid Methyltransferase from Rat Liver Microsomal Fraction, Biochem. J. 237: 699-705 (1986).

9.   D.E. Vance and B. de Kruijff, The Possible Functional Significance of Phosphatidylethanolamine Methylation, Nature, 288: 277-279 (1980).

10. R. Sundler and B. Akesson, Regulation of Phospholipid Biosynthesis in Isolated Rat Hepatocytes, Effect of Different Substrates, J. Biol. Chem., 250: 3359-3367

11. F. Makishima, S. Toyoshima and T. Osawa, Partial Purification and Characterization of Phospholipid N-Methyltransferases from Murine Thymocyte Microsomes, Arch. Biochem. Biophys. 238: 315-324 (1985).

12. D.E. Vance and N. Ridgway, Methylation of Phosphatidylethanolamine: Enzyme Characterization, Regulation and Physiological Function in Biological Methylation and Drug Design. R.T. Borchardt, C.R. Creveling and P.M. Ueland eds., Humana Press, Clifton, New Jersey (1986).

13. P.J. Brophy and D.E. Vance, Copurification of Choline Kinase and Ethanolamine Kinase from Rat Liver by Affinity Chromatography, FEBS Lett. 62: 123-125 (1976).

14. K. Ishidate, K. Nakagomi and Y. Nakazawa, Complete Purification of Choline Kinase from Rat Kidney and Preparation of Rabbit Antibody against Rat Kidney Choline Kinase, J. Biol. Chem. 259: 14706-14710 (1984).

15. K. Ishidate, K. Furusawa and Y. Nakazawa, Complete Co-Purification of Choline Kinase and Ethanolamine Kinase from Rat kidney and Immunological Evidence for Both Kinase Activities Residing on the Same Enzyme Protein(s) in Rat Tissues, Biochim Biophys. Acta 836: 119-124 (1985).

16. K. Tadokoro, K. Ishidate and Y. Nakazawa, Evidence for the Existence of Isozymes of Choline Kinase and Their Selective Induction in 3-Methylcholanthrene- or Carbon Tetrachloride-Treated Rat Liver, Biochim Biophys. Acta 835: 501-513 (1985).

17. P.C. Choy, P.H. Lim and D.E. Vance, Purification and Characterization of CTP: Cholinephosphate Cytidylyltransferase from Rat Liver Cytosol, J. Biol. Chem. 252: 7673-7677 (1977).

18. P.C. Choy and D.E. Vance, Lipid Requirements for Activation of CTP: Phosphocholine Cytidylyltransferase from Rat Liver, J. Biol. Chem. 253: 5163-5167 (1978).

19. D.E. Vance and P.C. Choy, Trends Biochem. Science 4: 145-148 (1979).

20. R. Sundler, G. Arvidson and B. Akesson, Pathways for the Incorporation of Choline into Rat Liver Phosphatidylcholines *In Vivo*, Biochim. Biophys. Acta 280: 559-568 (1972).

21. P.A. Weinhold, M.E. Rounsifer and D.A. Feldman, The Purification and Characterization of CTP: Phosphorylcholine Cytidylyltransferase from Rat Liver, J. Biol. Chem. 261: 5104-5110 (1986).

22.   D.E. Vance, P.C. Choy, S.B. Farren, P.H. Lim, W.J. Schneider, Asymmetry of Phospholipid Biosynthesis, Nature 270: 268-269 (1977).

23.   R.B. Cornell, K. Ishidate, N.D. Ridgeway, J.S. Sanghera and D.E. Vance, The Enzymes of Phosphatidylcholine Biosynthesis in "Enzymes of Phospholipid Metabolism, S. Gatt, ed., In Press (1987).

24.   P. Lim, R. Cornell and D.E. Vance, The supply of both CDP-Choline and Diacylglycerol can Regulate the Rate of Phosphatidylcholine Synthesis in HeLa Cells, Biochem. Cell Biol. 64: 692-698 (1986).

25.   S.L. Pelech, P.H. Pritchard and D.E. Vance, cAMP Analogues Inhibit Phosphatidylcholine Biosynthesis in Cultured Rat Hepatocytes, J. Biol. Chem. 256: 8283-8286 (1981).

26.   S.L. Pelech and D.E. Vance, Regulation of Rat Liver Cytosolic CTP: Phosphocholine Cytidylyltransferase by Phosphorylation and Dephosphorylation, J. Biol. Chem. 257: 14198-14202 (1982).

27.   S.L. Pelech, P.H. Pritchard, D.N. Brindley and D.E. Vance, Fatty Acids Promote Translocation of CTP: Phosphocholine Cytidylyltransferase to the Endoplasmic Reticulum and Stimulate Rat Hepatic Phosphatidylcholine Synthesis, J. Biol. Chem. 258: 6782-6788 (1983).

28.   P.H. Pritchard, P.K. Chiang, G.L. Cantoni and D.E. Vance, Inhibition of Phosphatidylethanolamine N-Methylation by 3-Deazaadenosine Stimulates the Synthesis of Phosphatidylcholine via the CDP-Choline Pathway, J. Biol. Chem. 257: 6362-6367 (1982).

29.   R. Sleight and C. Kent, Regulation of Phosphatidylcholine Biosynthesis in Cultured Chick Embryonic Muscle Treated with Phospholipase C, J. Biol. Chem. 255: 10644-10650 (1980).

30.   P.A. Weinhold, M.E. Rounsifer, S.E. Williams, P.G. Brubaker and D.A. Feldman, CTP: Phosphorylcholine Cytidylyltransferase in Rat Lung: The Effect of Free Fatty Acids on the Translocation of Activity Between Microsomes and Cytosol, J. Biol. Chem. 259: 10315-10321 (1984).

31.   P.H. Lim, P.H. Pritchard, H.B. Paddon and D.E. Vance, Stimulation of Hepatic Phosphatidylcholine Biosynthesis in Rats fed a High Cholesterol and Cholate Diet Correlates with Translocation of CTP: Phosphocholine Cytidylyltransferase from Cytosol to Microsomes, Biochim Biophys. Acta 753: 74-82 (183).

32.   S.L. Pelech, H.B. Paddon and D.E. Vance, Phorbol Esters Stimulate Phosphatidylcholine Biosynthesis by Translocation of CTP:Phosphocholine Cytidylyltransferase from Cytosol to Micrsomes, Biochim. Biophys. Acta 795: 447-451 (1984).

33.   S.L. Pelech, H.W. Cook, H.B. Paddon and D.E. Vance, Membrane-bound CTP: Phosphocholine Cytidylyltransferase Regulates the Rate of Phosphatidylcholine Synthesis in HeLa Cells treated with Unsaturated Fatty Acids, Biochim. Biophys. Acta 795: 433-440 (1984).

34.   D.E. Vance and S.L. Pelech, Enzyme Translocation in the Regulation of Phosphatidylcholine Biosynthesis, Trends Biochem. Science 9: 17-20 (1984).

35.   R.B. Cornell and D.E. Vance, Translocation of CTP:Phosphocholine Cytidylyltransferase from Cytosol to Membranes in HeLa Cells: Stimulation by Fatty Acid, Fatty Alcohol, Mono-and Di-acylglycerol, Biochim. Biophys. Act, In Press (1987).

36.   R.B. Cornell and D.E. Vance, Binding of CTP:Phosphocholine Cytidylyltransferase to Large Unilamellar Vesicles, Biochim. Biophys. Acta, In Press (1987).

37.   L.G. Herbette, C. Favreau, K. Segalman, C.A. Napolitano, and J. Watras, Mechanism of Fatty Acid Effects on Sarcoplasmic Reticulum II. Structural Changes Induced by Oleic and Palmitic Acids, J. Biol. Chem 259: 1325-1335 (1984).

PHOSPHOLIPIDS IN CELLULAR SURVIVAL AND GROWTH

J. K. Blusztajn, U. I. Richardson, M. Liscovitch
C. Mauron and Richard J. Wurtman

Department of Brain and Cognitive Sciences
Massachusetts Institute of Technology
Cambridge, Massachusetts

INTRODUCTION

Biological membranes (plasma membrane, nuclear envelope, endoplasmic reticulum, etc.) are composed primarily of phospholipids and proteins. Phospholipids are structural components; their physicochemical properties allow them to aggregate in aqueous environments to form lamellar bilayers, that are characteristic of biological membranes. This ability of phospholipids to spontaneously form noncovalently bound aggregates that can act as diffusion barriers (membranes) depends on the chemical composition of a given phospholipid mixture, and other biologically important membrane properties, like surface potential or microviscosity, are affected by this composition. It thus seems obvious that changes in membrane composition of mammalian cells ought to affect their ability to grow and even survive. However, the problem has not been studied much and we have no understanding of how phospholipids might contribute to regulation of cellular growth and rate of division. This chapter reviews evidence showing that various phospholipids are essential to cellular survival and growth. The special role of phosphatidylcholine in survival of cholinergic neurons, and changes in phospholipid turnover during the cell cycle are also briefly discussed.

PHOSPHATIDYLCHOLINE, PHOSPHATIDYLETHANOLAMINE AND PHOSPHATIDYLSERINE IN CELLULAR SURVIVAL AND GROWTH

Phosphatidylcholine

Phosphatidylcholine is the major lipid component of all eukaryotic cells comprising 30-60% of membrane phospholipids (Ansell et al., 1973). In 1955 Eagle showed that choline was necessary for the survival of human cell lines. However it was only recently demonstrated (by Esko and colleagues, 1980; 1981) that the choline was needed solely to serve as the metabolic precursor of phosphatidylcholine. They isolated a mutant of Chinese hamster ovary (CHO-K1) cells that was unable to survive at elevated temperature ($40^{o}C$) in the presence of choline. However, this mutant, called strain "58", grew when phosphatidylcholine was present in the medium. It was then demonstrated that mutant 58 had a thermolabile enzyme necessary for phosphatidylcholine synthesis, cytidinediphosphocholine synthetase (Kennedy and Weiss, 1956), which was not active at $40^{o}C$. Due to this defect the mutant lost half of its phosphatidylcholine contents during a 20 hour period, (which resulted in a decrease of phosphatidylcholine/phosphatidylethanolamine ratio from 1.4 to 0.6). Addition of phosphatidylcholine bypassed the defect, the lipid was incorporated into the cells and supported cellular growth. This explanation was further supported by evidence that spontaneous revertants were able to survive on choline alone, had normal phosphatidylcholine content, and normal cytidinediphosphocholine synthetase activity.

The studies of Esko and colleagues did not show whether it was the exact amount of cellular phosphatidylcholine that was important in determining cell growth, or whether it was the ratio of phosphatidylcholine to some other lipid that was regulatory. To answer this question we have investigated the effects of changing phosphatidylcholine content on cell growth, using a neuroblastoma x glioma hybrid cell line, NG108-15. Since these cells possess only a limited capacity to synthesize choline de novo, it was possible to alter their phosphatidylcholine content by varying choline concentrations in the serum-free growth media. Cells maintained in the absence of choline or in 0.2 μM choline began to degenerate by the fourth day in culture and did not survive beyond the eighth day. Their phosphatidylcholine content decreased from 3 nmol/10^4 cells to 2 nmol/10^4 cells by the fourth day. Cells that were grown in 2 μM choline maintained their phosphatidylcholine content and survived, but did not divide. Cells grown in 20 and 200 μM choline divided, such that by the eighth day in culture their number increased 4-5 fold and their phosphatidylcholine content was 5-6 nmol/10^4 cells. In order to investigate whether the total phosphatidylcholine content of the cells or its ratio to other phospholipids was responsible for these effects we grew the cells in the presence of 1 mM ethanolamine, which competitively inhibits phosphatidylcholine synthesis from choline (inhibitory constant, Ki=0.2 mM) and is the metabolic precursor of another membrane phospholipid, phosphatidylethanolamine. After four days in the presence of 0.2 μM choline plus 1 mM ethanolamine, the ratio of phosphatidylcholine to phosphatidylethanolamine in the cells decreased from 2 to 0.3 and the cells degenerated. The original ratio could be restored when 20 μM choline was added to the medium. This was also the lowest choline concentration that allowed cellular survival under these conditions: i.e. ten-fold higher than that required in the absence of ethanolamine.

These results show that variations in the concentrations of the phospholipid precursors in the growth media dramatically affect the amounts of these lipids in cells (up to several-fold). It is likely that the decrease in phosphatidylcholine content due to choline deprivation was responsible for cell death, whereas the increased amount of phosphatidylcholine, when extracellular choline was in abundance, allowed cellular proliferation in the serum-free media. The stimulatory effect of phosphatidylcholine on cell growth was offset by phosphatidylethanolamine, suggesting that the ratio of phosphatidylcholine to other phospholipids, rather than its absolute amounts, was important in determining the cellular survival and rate of growth.

Phosphatidylethanolamine

The effects of phosphatidylethanolamine on cell growth have also been investigated using the genetic approach of generating specific mutants that require special nutritional conditions for growth. One such strain, M.9.1.1, of the CHO-K1 line was isolated by Voelker and Frazier (1986). M.9.1.1 cells degenerate when grown in the absence of ethanolamine, lysophosphatidylethanolamine, or phosphatidylserine, all of which can serve as metabolic precursors of phosphatidylethanolamine. Under these conditions over a period of 24 hours the amount of phosphatidylcholine increased from 50 to 54% of total phospholipids whereas that of phosphatidylethanolamine and phosphatidylserine declined from 17 to 13% and 10 to 7%, respectively (i.e. the ratio of phosphatidylcholine to phosphatidylethanolamine increased from 2.9 to 4.1). It was determined that the M.9.1.1 cells have a decreased ability to synthesize phosphatidylserine which is the main precursor of phosphatidylethanolamine in the wild type CHO-K1 line. Thus the cells become dependent on other precursors when the amounts of phosphatidylserine synthesized in situ become limiting. It is interesting that the M.9.1.1 strain can apparently synthesize enough phosphatidylserine to support growth (see below) but not enough to support sufficient phosphatidylethanolamine synthesis. Similar results were obtained in a study by Kano-Sueoka, and colleagues (1983) in which various cell lines were screened for ethanolamine requirement. They found that for some cell lines (e.g. rat mammary carcinoma, 64-24; human mammary carcinoma, T-47D; mouse hybridoma, AS100-1) ethanolamine was a mitogen, whereas the rate of growth of other lines was not affected by ethanolamine (e.g. rat mammary carcinoma, WRK-1; rat hepatoma, HTC). Invariably, in the cells that required ethanolamine, phosphatidylethanolamine constituted 13-20% of phospholipids whereas in those that were independent of the precursor the same value was 30% (in these cells the ratio of phosphatidylcholine/phosphatidylethanolamine was 1.4-1.5; the same ratio in the cells requiring ethanolamine was 3-4.5). However when the cells that required ethanolamine were cultured in the presence of this compound, their phospholipid composition was similar to the other cell lines (i.e. 30% of phosphatidylethanolamine and 1.6-1.8 phosphatidylcholine/phosphatidylethanolamine ratio).

Phosphatidylserine

Phosphatidylserine is a minor component of biological membranes (Ansell et al., 1973), however due to its net negative charge it is probably important in determining membrane surface potential, and thus local (near membrane) pH, and ionic environment. Mutants of CHO-K1 cells that lack the capacity to synthesize this phospholipid (designated PSA-3) cannot survive unless their growth medium is supplemented with phosphatidylserine (Kuge et al., 1986). After two days without phosphatidylserine the amount of this lipid in PSA-3 cells decreased from 7 to 2.4% of total phospholipids. Interestingly other changes were also apparent: phosphatidylcholine content increased from 51 to 65% and that of phosphatidylethanolamine decreased from 18 to 9%, resulting in an increase in the ratio of these two lipids from 2.8 to 7, i.e. a ratio even higher than that observed in degenerating M.9.1.1 cell line (see above).

Conclusions

The data reviewed in this section suggest that phosphatidylcholine, phosphatidylethanolamine and phosphatidylserine are all necessary components of mammalian cells. Under normal conditions of growth the cells (at least in culture) contain fairly constant amounts of all of these lipids. Both genetic and nutritional manipulations may be used to alter these amounts. The resulting perturbations can be very dramatic: affecting both the phospholipid composition and cellular survival and growth rate. At present our understanding of how chemical structure of phospholipids affects their function within membranes is so rudimentary that we cannot explain how membrane phospholipid composition might affect cellular functions. It is possible to speculate that membranes that contain high proportions of phosphatidylethanolamine might be less stable [since this phospholipid tends to disrupt the bilayer by forming hexagonal II phases (Cullis et al., 1983)], however why reduced amounts of this lipid would be detrimental is not clear. The ratios of phosphatidylcholine to phosphatidylethanolamine were found to influence not only the ability of cells to survive, but also their rate of division (accelerating it in some cases, see above). What physicochemical parameters of phospholipid mixtures are important in exerting those effects, and what are the primary intracellular membranes whose functions are affected by changes in phospholipid composition remains to be elucidated.

PHOSPHOLIPID TURNOVER AND CELL CYCLE

The fact that phospholipids may influence the rate of cell growth implies that during the cell cycle the composition of cellular membranes might fluctuate. Changes in turnover of particular phospholipids might result, for example, in formation of membranes that are optimal for cell division or for increased rates of protein synthesis during the G1 phase

of the cycle. Surprisingly little is known about regulation
of phospholipid turnover in dividing cells. Cunningham (1972)
observed that when the mouse 3T3 cells stopped dividing by
forming a contact-inhibited monolayer their phospholipid
content did not change. However the turnover of
phosphatidylcholine was increased 2-fold, whereas that of
phosphatidylethanolamine and phosphatidylserine was decreased
by 85 and 48%, respectively. Initiation of cell division (by
addition of fresh serum to contact-inhibited cells) was
accompanied by a rapid increase in the rate of synthesis of
all phospholipids. Intracellular levels of phosphocholine and
phosphoethanolamine, precursors of the respective lipids, were
also found to increase upon initiation of cell division in a
similar study (Warden et al., 1980). The increase in
phosphocholine levels has been attributed to the activation of
choline kinase (Warden and Friedkin, 1985). In synchronized
mouse mastocytoma, P815Y, radiolabeled choline was
incorporated into phospholipids predominantly during the early
S and G2 phase (Lingwood and Thomas, 1975). Variations in
turnover of other phospholipids were not determined.

PHOSPHATIDYLCHOLINE IN CHOLINERGIC NEURONS

All cells use choline as the precursor of
phosphatidylcholine. Cholinergic neurons are unique, since
they alone utilize choline for an additional purpose,
synthesis of their neurotransmitter, acetylcholine. Therefore
the choline-phospholipids in cholinergic cells constitute a
large pool of choline that can potentially be used for
acetylcholine synthesis when the amounts of choline needed to
sustain acetylcholine release are enhanced (e.g. when
particular cholinergic neurons fire frequently, or when the
supply of choline from the extracellular fluid is inadequate).
The resulting depletion in the amount of choline-phospholipids
within cell membranes might be expected to alter membrane
functions, leading, perhaps, to pathological conditions when
membrane viability might be compromised at the expense of
maintained neurotransmission. Indeed a decrease in the amount
of phosphatidylcholine has been described in repeatedly
depolarized superior cervical ganglia (Parducz, et al., 1976).
To investigate this process in some detail we have used slices
prepared from a region of the brain that contains high numbers
of cholinergic neurons, the striatum. When rat striatal
slices superfused with a choline-free medium containing
eserine were stimulated continuously for 30 minutes, they
released acetylcholine at a constant high rate; however,
tissue acetylcholine and choline levels failed to decline
significantly (Maire and Wurtman, 1985). Furthermore,
addition to the medium of hemicholinium-3, a drug that blocks
the high-affinity uptake of choline into cholinergic synaptic
boutons, terminated acetylcholine release. It was thus
suggested that a major source of the choline that had been
used for acetylcholine synthesis by the slices was the
phosphatidylcholine in synaptic membranes (Maire and Wurtman,
1985): When the neurons were depolarized, some of their phosphatidyl-

choline would be hydrolyzed, causing choline to be released into the extracellular fluid and then taken up into cholinergic terminals for conversion to acetylcholine. We therefore measured the amounts of phospholipids present in the slices prior to, and after the period of electrical stimulation, and found that it declined by about 22%. In confirmation of previous findings (Maire and Wurtman, 1985), addition of choline in physiologic concentrations (20 μM) to the medium markedly enhanced acetylcholine release, both basally and during stimulation. Moreover, the choline completely protected the slices from the phospholipid depletion observed in its absence. Tetrodotoxin (a drug that inhibits neuronal firing) blocked both the stimulation-induced release of acetycholine and the concurrent depletion of phospholipids, suggesting that this depletion was contingent upon neuronal depolarization (and, perhaps, the use of the additional "free" choline for acetylcholine synthesis).

The phospholipids in brain membranes are rich in phosphatidylethanolamine and phosphatidylserine as well as phosphatidylcholine; together, these phosphatides account for 85% of total brain phospholipids (Ansell et al., 1973). Moreover, the choline in phosphatidylcholine represents about 80% of the total membrane-bound choline in the brain (Ansell et al., 1973), suggesting that phosphatidylcholine would be the most likely cellular reservoir for free choline that could be used for acetylcholine synthesis. To determine whether the reduction in total brain phospholipids that we observed in stimulated slices superfused without choline represented phosphatidylcholine exclusively, or also included other brain phospholipids, we fractionated and quantified these compounds in some of the tissues obtained before and after electrical stimulation, and also expressed tissue phosphatidylcholine as a percent of total phosphatidylcholine, phosphatidylethanolamine, and phosphatidylserine. Stimulation in the choline-free medium caused a 15 percent decline in membrane phosphatidylcholine, but also led to proportionate decreases in phosphatidylethanolamine and phosphatidylserine, so that phosphatidylcholine as a percent of the three main structural phospholipids did not decline. In three separate perfusion studies the ratio of phosphatidylcholine/(phosphatidylcholine+phosphatidylethanolamine+phosphatidylserine) was 0.42 prior to stimulation and 0.43 subsequent to stimulation. Addition of choline to the superfusion medium also failed to affect this ratio (which was 0.37 both initially and after the stimulation).

These data support the hypothesis (Blusztajn and Wurtman, 1983) that membrane phosphatidylcholine can serve as a reservoir of choline to be used for acetylcholine synthesis. When neurons are provided with adequate free choline and stimulated electrically, they produce and release acetylcholine without depleting this reservoir; however, when the precursor is otherwise unavailable, the net catabolism of the phosphatide to liberate free choline is sufficient to reduce membrane levels of phosphatidylcholine and other phosphatides significantly. The choline thus liberated could be taken up and converted to acetylcholine in adjacent cholinergic nerve terminals, and perhaps also used for phospholipid resynthesis in all cells.

Interestingly, the depolarization-induced depletion of phosphatidylcholine appears not to be specific for this phosphatide since the reduction in its levels was matched by proportionate reductions in the other two structural phospholipids. A prolonged imbalance between the amounts of free choline that are available to cholinergic neurons and the amounts needed to sustain acetylcholine release might thus be expected to interfere with the cells' ability to produce and remodel its membranes, with consequent changes in its functional properties and, perhaps, survival (Blusztajn and Wurtman, 1983).

Striatal slices contain only a small proportion of cholinergic terminals, and therefore it is possible that the reduction in their phospholipids might have occurred also within noncholinergic neurons, and that the choline liberated from those cells provided the choline needed to sustain acetylcholine synthesis. We therefore studied (Blusztajn et al., submitted) the transfer of choline form phosphatidylcholine to acetylcholine in purely cholinergic population of cells, a human neuroblastoma cell line, LA-N-2 (West et al., 1977). The cells were grown in the presence of [^{3}H-methyl]methionine which is intracellularly converted to radiolabeled S-adenosylmethionine. This compound is a methyl donor for the only pathway of de novo choline biosynthesis, which is catalyzed by an enzyme, phosphatidylethanolamine N-methyltransferase (Bremer et al., 1960). The product of this reaction is phosphatidylcholine.

The amount of labeled phosphatidylcholine synthesized in LA-N-2 cells (preincubated with dimethylethanolamine to increase the rate of phosphatidylcholine synthesis) during a 20 hour incubation period in the presence of [^{3}H-methyl] methionine was 0.43 nmol/mg protein. This constituted 0.6% of the total cellular phosphatidylcholine. The amount of radioactivity found in choline phosphocholine and acetylcholine consituted 0.1, 0.3 and 0.3% respectively of that found in labeled phosphatidylcholine. Although the free choline pool was rather small (0.13 nmol/mg protein) its specific radioactivity was half (i.e. 6 dpm/pmol) that of phosphatidylcholine, indicating that posphatidylcholine was indeed the precursor of Ch. These data suggest that a large proportion of free choline was derived from phosphatidylcholine, and that the newly - formed phosphatidylcholine probably entered the bulk pool of phospholipids. The amount of acetylcholine in the cells was fourteen fold larger than that of choline (1.9 nmol/mg protein); however the specific rafioactivity of acetylcholine was only one tenth that of phosphatidylcholine (or one fifth that of Ch), suggesting that most of the cellular acetylcholine might have been present in a stable compartment that did not turn over during the labeling period. Assuming that the labeled acetylcholine was synthesized from choline whose specific radioactivity was 6 dpm/pmol, the total amount of newly-formed acetylcholine would have constituted 20% of its pool.

These data indicate that cholinergic neurons form phosphatidylcholine by methylating phosphatidylethanolamine, and that this phosphatidylcholine can be hydrolyzed to free choline, which can then be acetylated to form acetylcholine. Preliminary observations suggest that this acetylcholine enters a releasable pool, inasmuch as radiolabeled acetylcholine was recovered from the growth media of LA-N-2 cells.

It remains to be determined whether the choline used for acetylcholine synthesis is derived from a specific phosphatidylcholine pool (e.g. one synthesized by phosphatidylethanolamine N-methyltransferase; or localized in a specific organelle like synaptic vesicles (Parducz et al., 1976); or one containing a particular combination of fatty acids; or from the general catabolism of bulk phosphatidylcholine. In the former case, it is likely that a regulatory mechanism exists that can control the relative rates of phosphatidylcholine synthesis and/or degradation, depending on the requirements of choline for acetylcholine synthesis. This regulatory mechanism may be influenced by neuronal firing. When more choline is needed for acetylcholine synthesis (during periods of frequent firing) the rate of phosphatidylcholine hydrolysis might be increased. Thus the actual contribution of phosphatidylcholine hydrolysis in providing choline for acetylcholine synthesis may vary according to acetylcholine output and exogenous choline supply.

CONCLUSIONS

The significance of phospholipids in maintaining cellular viability has been now well documented. It appears that the proportions of various phospholipids within the cellular membranes are more important than the actual amounts of each of the phospholipid class. It is likely that these proportions are regulated during the cell cycle.

Our understanding of how the physicochemical properties of phospholipids contribute to membrane structure and function is inadequate to explain why the changes in phospholipid composition of cells affect cellular survival and the rate of growth.

Cholinergic neurons use phosphatidylcholine in their membranes as a storage pool of choline for acetylcholine synthesis. When the demand for choline transcends the rate of phosphatidylcholine resynthesis the composition or amount of membrane might undergo changes that adversely affect neuronal viability.

REFERENCES

Ansell, G. B., Hawthorne, J. N., and Dawson, R. M. C., 1973,

"Form and Function of Phospholipids," Elsevier, Amsterdam.

Blusztajn, J. K., and Wurtman, R. J., 1983, Choline and cholinergic neurons, Science, 221:614.

Blusztajn, J. K., Liscovitch, M., and Richardson, U. I., submitted, Synthesis of acetylcholine from choline derived from phosphatidylcholine in a human neuroblastoma cell line, LA-N-2.

Bremer, J., Figard, P. H. and Greenberg, D. M., 1960, The biosynthesis of choline and its relation to phospholipid metabolism, Biochim. Biophys. Acta, 43:477.

Cullis, P. R., de Kruijff, B., Hope, M. J., Verkleij, A. J., Nayar, R., Farren, S. B., Tilcock, C., Madden, T. D., and Bally, M. B., 1983, Structural properties of lipids and their functional roles in biological membranes, in: "Membrane Fluidity in Biology" vol. 1, Academic Press, New York.

Cunningham, D. D., 1972, Changes in phospholipid turnover following growth of 3T3 mouse cells to confluency, J. Biol. Chem., 247:2464.

Eagle, H., 1955, The minimum vitamin requirements of the L and HeLa cells in tissue culture, the production of specific vitamin deficiencies, and their cure, J. Exptl. Med., 102:595.

Esko, J. D., and Raetz, C. R. H., 1980, Autoradiographic detection of animal cell membrane mutants altered in phosphatidylcholine synthesis, Proc. Natl. Acad. Sci. USA, 77:5192.

Esko, J. D., Wermuth, M. M., and Raetz, C. R. H., 1981, Thermolabile CDP-choline synthetase in a animal cell mutant defective in lecithin formation, J. Biol. Chem., 256:7388.

Kano-Sueoka, T., Errick, J. E., King, D., and Walsh, L. A., 1983, Phosphatidylethanolamine synthesis in ethanolamine-responsive and -nonresponsive cells in culture, J. Cell. Physiol., 117:109.

Kennedy, E. P., and Weiss, S. B., 1956, The function of cytidine coenzymes in the biosynthesis of phospholipids, J. Biol. Chem., 222:193.

Kuge, O., Nishijima, N., and Akamatsu, Y., 1986, Phosphatidylserine biosyntheses in cultured Chinese hamster ovary cells. II. Isolation and characterization of phosphatidylserine auxotrophs. J. Biol. Chem., 261:5790.

Lingwood, C. A., and Thomas, D. B., 1975, Modulation in the rates of incorporation of lipid precursors during cell cycle, J. Cell. Physiol., 86:635.

Maire, J-C., an Wurtman, R. J., 1985, Effects of electrical stimulation and choline availability on the release and contents of acetylcholine and choline in superfused slices from rat striatum, J. Physiol., Paris, 80:189.

Parducz, A., Kiss, Z., and Joo, F., 1976, Changes of phosphatidylcholine content and the number of synaptic vesicles in relation to neurohumoral transmission in sympathetic ganglia, Experientia, 32:1520.

Warden, C. H., Friedkin, M., and Geiger, P. J., 1980, Acid-soluble precursors and derivatives of phospholipids increase after stimulation of quiescent Swiss 3T3 mouse fibroblasts with serum, Biochem. Biophys. Res. Commun., 94:690.

Warden, C. H., and Friedkin, M., 1985, Regulation of choline kinase activity and phosphatidylcholine biosynthesis by mitogenic growth factors in 3T3 fibroblasts, <u>J. Biol. Chem.</u>, 260:6006.

West, G. J., Uki, J., Herschman, H. R. and Seeger, R. C., 1977, Adrenergic, cholinergic, and inactive human neuroblastoma cell lines with the action potential Na^+ ionophore, <u>Cancer Res.</u>, 37:1372.

Wurtman, R. J., Mauron, C., and Blusztajn, J. K., submitted, Choline protects against depolarization-induced decreases in brain phosphatide levels.

Voelker, D. R., and Frazier, J. L., 1986, Isolation and characterization of a Chinese hamster ovary cell line requiring ethanolamine or phosphatidylserine for growth and exhibiting defective phosphatidylserine synthase activity, <u>J. Biol. Chem.</u>, 261:1002.

NUTRITIONAL CONSIDERATIONS OF LECITHIN ADMINISTRATION

W. Feldheim

Department of Human Nutrition and Food Science
University of Kiel
Düsternbrooker Weg 17-19, D 2300 Kiel 1, FR Germany

INTRODUCTION

The aim of this presentation is to evaluate the significance of Lecithin-Choline as NUTRIENTS in our diet.

In general, the position and importance of the different constituents of our foodstuffs are judged by their physiological properties and functions.

As potential sources for energy, the position of fats, carbohydrates, alcohol and in case of emergency of proteins is well known. Some of the metabolites of these food components are used, in addition, for the supply with carbon or nitrogen compounds, of anabolic reactions of the metabolism. In relation to the requirements of the body, the greatest part of the food eaten (except water) consists of those macronutrients, essential for life, necessary for energy metabolism.

Being essential in another way, the function and position of the micronutrients of our food is quite different. Contrary to the first group of energy providing organic compounds, the average intake necessary for those components is very small. The possibility of a biosynthesis in the body does not exist, or is very limited or insufficient in periods of a higher need. On the other hand, there is a requirement for a supply in connection with some function for those nutrients, e.g. in metabolism or cell structure. Among those components, vitamins, essential amino acids and fatty acids, but also inorganic matter, such as minerals and trace elements are found. The secret of an appropriate nutrition behaviour is to eat a mixture of different foods, providing the body with a sufficient amount of those micronutrients and just as much metabolizable energy, as is necessary to keep the body weight of an adult approximately constant.

Among these schedules of classification, it is not easy to find the position of lecithins. Looking at their general structure as a whole, the relationship between fats and lecithins is obvious, but the special structure at the carbon number 3 of the glycerol-part and the relatively small amount of lecithins in food is something particular among the fats (Figure 1).

FAT CONTENT IN DIET

In different countries, the fat content of the diets expressed as energy covers a range between a very low amount (a few percent of total energy in developing countries) and up to 50% or more in the Western world. This high amount of fats in the diet is responsible for a high incidence of overnutrition and obesity in these countries. It is recommended to decrease the fatty component of the diets to 30 or 25% of the total energy. In practice, this means a reduction of daily fat from 130 - 150g to 60-80g for an adult.

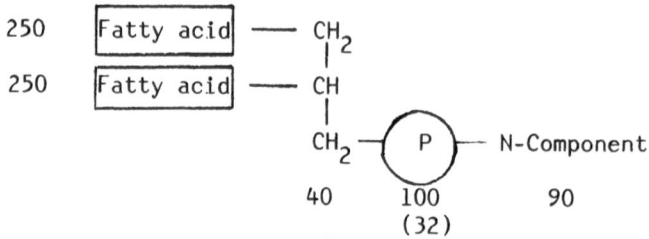

Fig.1: General Formula of Lecithins

In this situation the participation of lecithin is insignificant due to their small intake. With the decreasing amounts of energy and foods it seems necessary to prevent a further reduction of lecithin intake.

FATTY ACID COMPOSITION IN LIPIDS

Instead of saturated fatty acids, more unsaturated fatty acids should be preferred in a healthy diet. It is well known that the members of the polyunsaturated fatty acids (PUFA) have great significance as precursors of prostaglandins, acting as regulators of many important metabolic ways. Lecithins having PUFA or very high PUFA (perhaps in the phospholipids of oils of marine origin) in their structure, may therefore be of considerable interest for future research. The supply of PUFA, needed in the range of 1-2g daily, and a recommended daily allowance of 5-7g, is not a problem if a normal diet is consumed. But with decreased amounts of fats consumed as recommended, an advantage could be seen in changing the pattern of the fatty acids with the help of lecithins containing essential fatty acids (Figure 2).

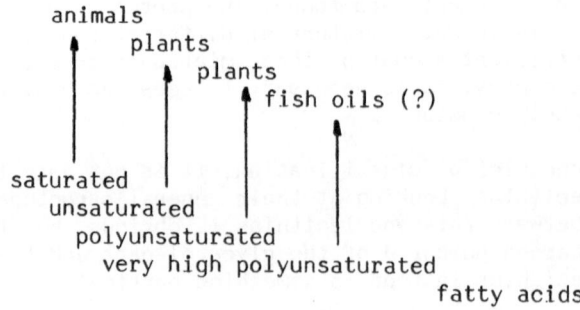

Fig. 2: Fatty Acid Components of Lecithins

PHOSPHORUS

Phosphorus is an important inorganic constituent of nutrition. Phosphorus is used as an inorganic part of the skeletal matrix and, in the soluble organic forms, as a component of the energy metabolism, and in membranes as well. A typical Western diet contains many different phosphorus sources. The regulation of the absorption of phosphorus in the intestine is extended to a wide range of the different phosphorus concentrations of the diet. The intake of phosphorus is absorbed mainly as inorganic phosphate, and the excess is excreted, so the metabolism is balanced. At low concentrations, there is an active transport for phosphorus from the intestine, and the excretion of phosphate falls markedly.

The phosphorus-intake from lecithins is only minor, and therefore is considered to have a small influence on the phosphorus-supply.

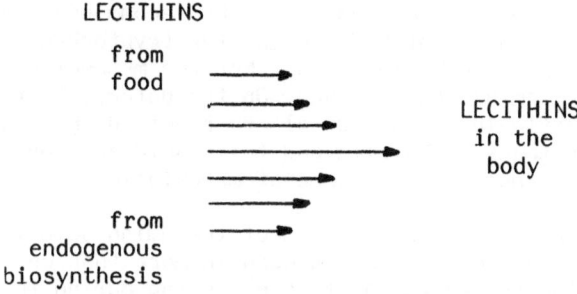

Fig. 3: Sources of Lecithins

ETHANOLAMINE MOIETY OF LECITHIN

Coming now to the ethanolamine-derived part of the molecule, methylated to a different degree, the situation seems to be similar to that of the other constituents of lecithins. This N-containing part can be synthesized in the body, totally or partly. The total of the body lecithins is delivered by the lecithins from foodstuffs and the de novo formed molecule. The situation is comparable with the supply of cholesterol, coming from exogenous and endogenous sources as well (Figure 3).

But, concerning the part of the choline, it must be considered that, during the biosynthesis, the methylation step is dependent on the availability of another essential component, available to a limited extent, the amino acid methionine. In this situation, it seems reasonable to deliver lecithins in a sufficient amount with the daily food.

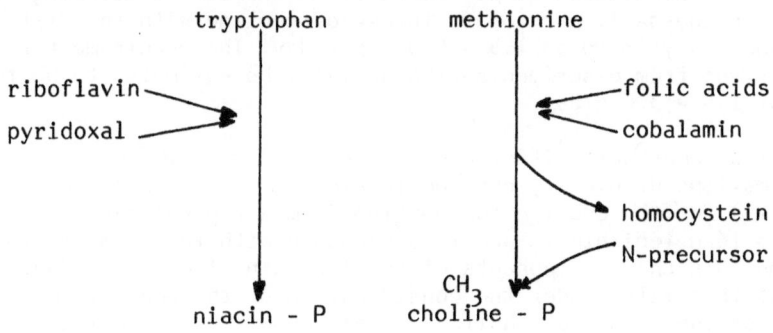

Fig. 4: Pathways for Niacin and Choline Formation in the Body

In this respect, there are parallels to the vitamin niacin (nicotinic acid) and the lecithin-choline. This essential component is part of our food, so portions of the requirement are covered with the diet. In the metabolism, the conversion of the amino acid tryptophan to form nicotinic acid is possible. This reaction takes place at different extent, depending on the needs of the metabolism. The conversion factor of 60 is an approximation, based on a limited number of studies. In the stages of pregnancy, more niacin is formed from tryptophan. The average intake of niacin (in form of equivalents) is about 15 to 20 mg per day.

The calculated amount of the mean daily consumption of lecithin in Western Europe is about 1.5g, and (according to the structure formula) 10% = 150mg thereof are forming the N-containing fraction of lecithin.

Choline is assumed to be 25 % of those components and calculated to be about 40mg at this lecithin intake.

For both the amino acids compared here, the recommended daily allowances are 3 mg/per day and kg of body weight for tryptophan and 10 mg for the total of S-containing amino acids. For the metabolism of both amino acids, other vitamins are involved. On the pathway to form niacin from tryptophan, riboflavin and pyridoxal are required. For the homocystein-methionine pathway, cobalamin and folic acid are involved. There is an influence of retinol on the degree of methylation.

In both cases, the discussed function of the amino acid is only one among others and the formed compound is used in more than one function. In the case of methionine, we found it to be acting during the methylation of mono- and dimethyltocols (non-alpha-tocopherols) to form trimethyl-tocol (alpha-tocopherol) at the germination of the seeds and the first stages of development of most plants (Feldheim, 1986). Both conversion products, lecithin-choline and alpha-tocopherol, are parts of the membranes. In this respect, the "weak" position of tocopherols among the vitamins is another remarkable point, when seen in relation to the same position of choline.

The most powerful agent and key metabolite in transmethylation in general is the S-adenosylmethionine, used in different reactions of the C_1-metabolism. For some unknown reasons, the methylation of ethanolamine to form choline is the most vulnerable point in case of a shortage of methyl donors. For that other reason, a sufficient intake of lecithin-choline with the diet is recommended.

SUMMARY AND CONCLUSIONS

According to the Recommended Dietary Allowances (1974), mixed diets are estimated to provide human adults with 400-900 mg of choline daily. Such amounts are considered to be adequate. At the same time, Lang (1973) has calculated the daily per capita intake of choline with the diet based on food analysis to be 1.5 - 4.0 g per day. The requirement for choline, derived from experiments with animals, he estimated to be in the range of 1.5 - 3.0 g/day.

There is a remarkable difference between earlier and more recent estimations of dietary choline intake. In the calculation of Wurtman (1979) for the choline consumption from a typical day's diet, based on a 3.16 g lecithin intake in accordance with the formula, 10 %, about 300 mg, are the N-components of the lecithin. There is a further reduction of this value under the consideration of the non-choline-components, so the amount of choline is only about 100 mg. Using the data of Cairella (1983) for Italy, lecithin consumption amounts to an average of 1.58 g or about the half of the value, seen by Wurtman (1979).

Here the same calculation schedule leads to an amount of choline of 50 mg only.

The study presented here by Åkesson and Nilsson has shown that the intake of lecithin is about 2 mmol per day with a normal adult's diet and 1.5 mmol with a vegetarian diet. Concerning the recommendation to eat less meat and eggs but more fruit and vegetables, the estimated choline is about 50 mg , which is in good agreement with the newer calculations.

From these few examples concerning intake and estimated requirements of lecithin-choline a great uncertainty is seen. Perhaps the earlier calculated values, based on insufficient analytical procedures and unspecified determinations, are too high. Therefore, all the data, used in the calculation and interpretation of the lecithin and choline content of our foodstuffs must be reviewed critically, under special consideration of the more recent investigations. A further complication may be the fact that other N-components of the lecithins are used as precursors of choline in metabolism at least partly. But, up to this point of clarification, a plentiful supply of dietary lecithins is recommended.

REFERENCES

Åkesson, B. and Nilsson, A., 1987, This Book

Bhuteni, V., Sharma, Y.K. and U.K. Misra, 1986: Vitamin A and Phospholipid Methylation in various Organs of Rats. Nutr.Rep. Intern. 34: 129

Cairella, M., Dann Treves, L., Del Balzo, V., Godi, R., Pretaroli, A.R., Scatena, R., and Scherer, R., 1985, Lecithin Consumption in the Western European Diet in F. Paltauf and D. Lekim eds., Lecithin and Health Care, Semmelweis-Verlag, Hoya

Feldheim, W., Lausch, G., Schulz, H., and Cummings, P.H., 1986, Effect of Germination on Tocopherol Pattern of the Seeds of Wheat, Lupin, Soya and Sunflower. Nutrition 10 : 21

Lang, K., 1973, Biochemie der Ernährung, Steinkopff-Verlag, Darmstadt

National Academy of Sciences, 1974, Recommended Dietary Allowances, Washington, D.C.

Wurtman, J.J., 1979, Sources of Choline and Lecithin in the Diet, Nutrition and the Brain, A. Barbeau, J.H. Growdon and R.J. Wurtman, eds. Raven Press, New York

Zeisel, S.H., 1985, in Lecithin in Health and Disease, B.F. Szuhaj and G.R. List, eds., Am. Oil Chem. Soc.

USE OF SOYBEAN LECITHIN IN PARENTERAL NUTRITION

Guy Dutot

Laboratoires Cernep Synthelabo R & D
Le Plessis Robinson, France

INTRODUCTION

Total parenteral nutrition is indicated for patients who cannot be adequately fed enterally and who are nutritionally depleted. These patients usually fit into the following categories : gastrointestinal disorders, malnutrition, short bowel syndrome, post operative patients, cancer, burns, trauma.

Total parenteral nutrition should provide all the following nutritional requirements :

- calories : carbohydrates and lipids
- nitrogen : amino acids and protein hydrolysates
- vitamins
- minerals
- trace elements

Lipids are used as a source of essential fatty acids to prevent or treat essential fatty acid deficiency or as an energy source.

The high caloric value (9 kcal/g) and the low osmolality of lipids allows the administration of a high energy content in a small volume via peripheral veins.

PHYSICO-CHEMICAL PROPERTIES OF AN INJECTABLE FAT EMULSION

For intravenous administration, lipid emulsions, as injectable grade products, should have the following characteristics :

- sterile
- pyrogen-free
- isotonic to blood plasma
- neutral : pH between 6 and 8
- physically stable.
- mean particle size around 300 nm and no droplet over 5 microns to minimize the risk of pulmonary embolism.

The usual composition of an injectable fat emulsion is presented in Table 1.

Table 1 : Composition of a Fat Emulsion

Triglycerides (10 % to 20 %)	→	purified vegetable oils
Emulsifier (0.7 % to 1.5 %)	→	soybean or egg yolk phosphatides
Isotonicizer (2.25 %)	→	glycerol
Water for injection		

Natural phospholipids are used as emulsifiers because it is well documented that synthetic emulsifiers are responsible for undesirable side effects when infused in man.

INFLUENCE OF PHOSPHOLIPIDS ON LIPID EMULSION METABOLISM

When it enters the bloodstream, the lipid droplet is rapidly modified as it acquires apoproteins (Table 2). Triglycerides are subsequently hydrolyzed to free fatty acids and glycerol by lipoprotein lipase. Fatty acids will be either oxidized to meet the energy requirements of the patient or reesterified to triglycerides and stored in adipose tissue or secreted by the liver

Table 2 : Metabolism of Intravenous Fat

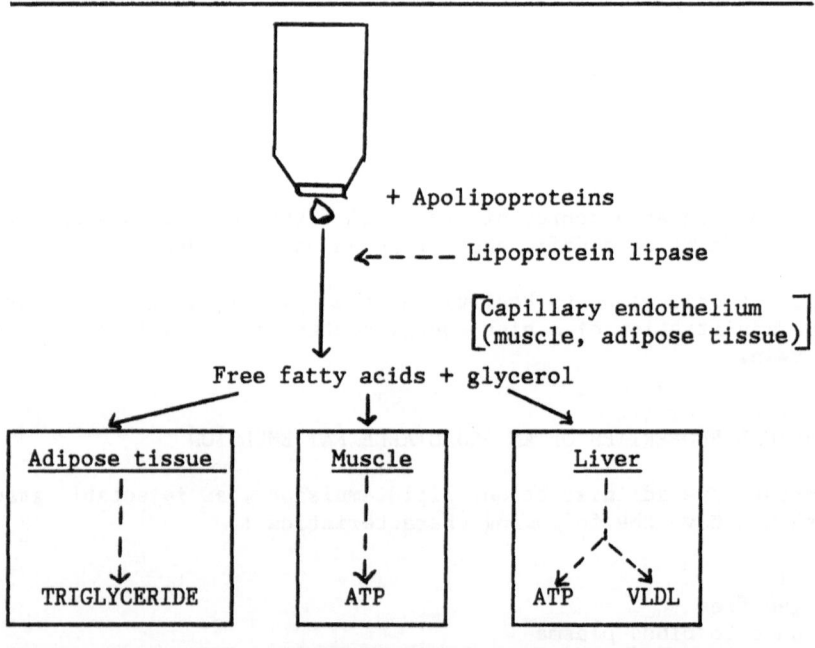

in the form of lipoproteins. The location of phospholipids outside the fat globule explains that it can influence the fate of lipid emulsion. Indeed, phospholipids stabilize the emulsion by adsorbing at the oil-water interface. The two fatty acyl chains are directed towards the oil phase while the hydrophilic moiety is located in the water phase. Numerous studies have shown that the nature, quality and composition of the emulsifier influence the stability, tolerance and metabolism of the fat emulsion.

Pharmacokinetics of lipid emulsions

Influence of the nature of emulsifier. Fat emulsion clearance from the bloodstream must be checked very carefully as the dosage will be adjusted to the patient's ability to metabolize lipids in order to avoid lipid accumulation in plasma.

Ziak and coworkers[1] have studied the kinetic properties of three soybean oil emulsions stabilized with egg phosphatides (Intralipid®), soybean phosphatides (Lipofundin®) or hydrogenated soybean phosphatides (Venolipid®). The emulsions were infused in six healthy volunteers at a dosage of 0.1 g/kg/h for 8 hours. They found that Venolipid® (with hydrogenated phosphatides) was cleared at a much lower rate than Intralipid® or Lipofundin®. While serum triglycerides return to normal values 3 hours after the end of infusion in Intralipid and Lipofundin-treated patients, their concentration is three times the fasting value in the Venolipid group.

The same experiment was conducted by Forster et al. (1979)[2] with Intralipid® and Lipofundin® using a higher dosage (0.25 g/kg/h). These investigators noted that Intralipid®, stabilized with egg yolk lecithin, is eliminated much more slowly than Lipofundin® containing soybean lecithin.

Effects of oil-phospholipid ratio. Intravenous lipid emulsions are generally used at 10 % or 20 % oil concentration stabilized with the same amount of phosphatides. The different oil-phosphatides ratios resulting from such compositions may affect the fat emulsion clearance.

The effects of long term administration of two soybean oil emulsions containing either 10 % or 20 % oil stabilized with 1.2 % egg yolk phosphatides have been studied by Izzo et al.[3] in beagle dogs. Both emulsions were infused at a dosage of 4 g/kg/day for 91 days. Consequently, animals infused with the 10 % emulsion receved twice as much phospholipids as those infused with the 20 % fat emulsion. Serum triglycerides, phospholipids and cholesterol were measured at weekly intervals during the study. Although both groups were provided with the same amount of oil, the increase in serum triglyceride level was much more important in animals receiving the higher dosage of phospholipids. These results led to the hypothesis that the phospholipid film covering the fat globule could interfere with triglyceride metabolism. A significant rise in serum phospholipids occurred in animals infused with the 10 % emulsion. The most striking observation was an increase in cholesterol values closely related to variations in phospholipids concentrations.

The modifications in circulating cholesterol can be attributed to an increased mobilization from various tissues by exogenous phospholipids. The excess of circulating phospholipids and cholesterol is responsible for the appearance of lipoprotein X (Table 3), as described in patients receiving intravenous fat emulsions[4].

Table 3 : Mechanism of Lipoprotein X Appearance in Plasma of Patients
Receiving Intravenous Fat Emulsion

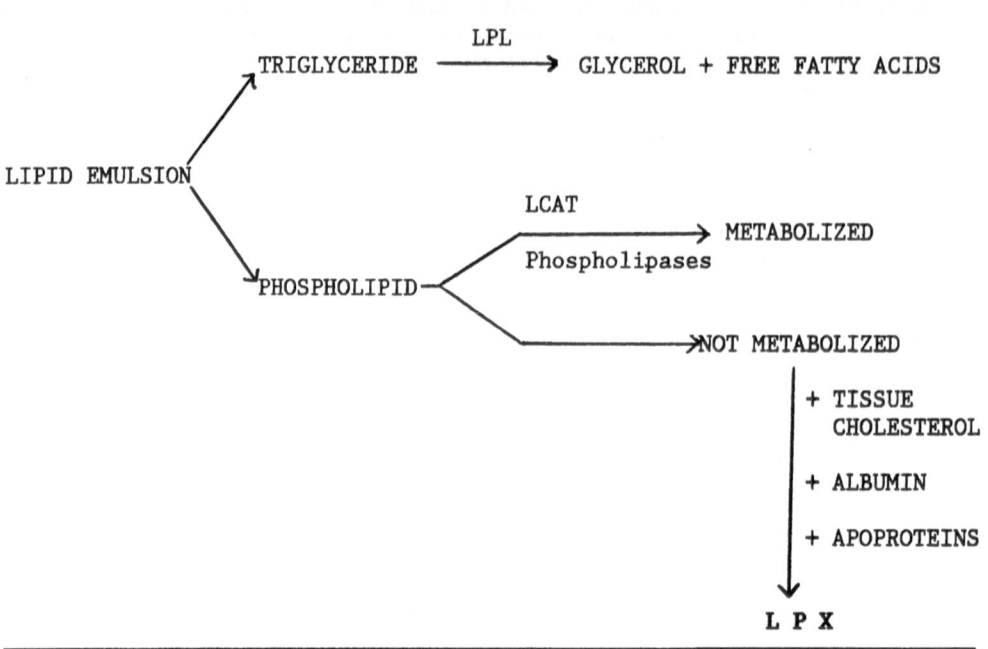

Fat emulsion creaming

Various factors can destabilize the lipid emulsion which may result in
breaking, coalescence or creaming of fat droplets.

When a lipid emulsion is incubated with serum in vitro, the mixture
normally remains as a uniform opalescent emulsion. Creaming is characterized
by the formation of a definite layer of emulsion above a clear serum layer.
Fat emulsion creaming was first reported in 1961 by Le Veen and coinvestigators[5], who observed that 70 % of patients infused with lipids were able to
cream the fat emulsion in vitro. These patients, called creamers, removed infused fat more rapidly than non creamers. Forbes[6] suggested that a creaming
phenomenon could be at least partly responsible for a rapid removal of fat
particles by the reticulo endothelial system in these acutely ill patients.
More recently, Hulman et al.[7] have tested the ability of sera obtained in
different groups of patients to cream a lipid emulsion. They observed that
about 90 % of intensive care patients were creamers while there were none in
the control group. Levels of serum C-reactive protein, a protein known to increase in association with inflammation, were measured and found to be
closely related to the degree of creaming.

The influence of C-reactive protein on fat emulsion creaming can be
explained by its interactions with lipids. C-reactive protein has a calcium
dependent binding specificity for phosphorylcholine. Consequently, it binds
to any phospholipid containing this structure, such as phosphatidylcholine
and sphingomyelin.

Dutot et al.[8] have studied the stability against creaming of two
emulsions containing egg yolk or soybean phosphatides. They found that
soybean phosphatides are less sensitive to creaming than egg yolk

phosphatides. Therefore, it appears that the nature of phospholipids can influence the degree of creaming. So far, no fat emulsion creaming in vivo has been reported. However, if it occurs, it could certainly modify the metabolism of lipid emulsion.

Effects of minor lecithin components on arterial blood pressure

Haemodynamic effects have been reported in animals infused with lecithin. Schubert and Wretlind[9], injecting lecithin in cat, observed a fall in blood pressure and apnea.

Tokumura et al.[10], in 1978, isolated in soybean lecithin a fraction with a depressor activity. This fraction was identified to be lysophosphatidic acid, a very potent depressive agent (0.1 % is sufficient to cause hypotension in the cat). Haemodynamic effects can be suppressed by fractionation of lecithin[11]. Table 4 shows the influence of lecithin purification on depressor response obtained in cats infused with egg or soybean lecithin. Haemodynamic activity is completely abolished by a previous fractionation of lecithin.

Table 4: Decrease in Arterial Blood Pressure Values (in mm Hg)

SOYBEAN PHOSPHATIDES		EGG YOLK PHOSPHATIDES	
NON FRACTIONATED	FRACTIONATED	NON FRACTIONATED	FRACTIONATED
- 67.5 ± 15.0	0.0	- 67.7 ± 21.5	- 7.3 ± 12

CONCLUSIONS

It is evident that emulsifier has a great influence on lipid emulsion metabolism and tolerance.

An injectable-grade lecithin requires careful selection and purification of the raw material, and a thorough knowledge of the side-effects of this emulsifier.

REFERENCES

1. E. Ziak, H. Pristautz, D. Brandt, K. Schaupp, and H.E. Musil, Comparison of three fat emulsions for parenteral nutrition, Ernahrung/Nutrition, 8:617 (1984).
2. H. Forster, R. Quadbeck, and A. Anschutz, Untersuchungen zur frage der dosierung und zur bedeutung von fett bei parenteraler ernährung, Infusionstherapie, 6:362 (1979).
3. R.S. Izzo, S. Larcker, W. Remis, J. Mennear, E. Woods, and N. Leissing, The effects on beagles of long term administration of 20 % Travamulsion fat emulsion, JPEN, 8:160 (1984)
4. D. Rigaud, P. Serog, A. Legrand, M. Cerf, M. Apfelbaum, and S. Bonfils, Quantification of lipoprotein X and its relationship to plasma lipid profile during different types of parenteral nutrition, JPEN, 8:529 (1984).
5. H. H. Le Veen, P. Giordano, and J. Spletzer, The mechanism of removal of intravenously injected fat, Arch. Surg., 83:169 (1961).

6. G. B. Forbes, Splenic lipidosis after administration of intravenous fat emulsions, J. Clin. Pathol., 31:765 (1978).

7. G. Hulman, I. Fraser, H. J. Pearson, and P. R. F. Bell, Agglutination of Intralipid by sera of acutely ill patients, Lancet, 8313:1426 (1982).

8. G. Dutot, I. Collignon, M. Pays, and J. Y. Wessely, Influence de la nature des phosphatides sur le crémage in vitro des émulsions lipidiques induit par les sérums humains riches en CRP, Gastroenterol. Clin. Biol., 8:77 (1984).

9. O. Schubert, and A. Wretlind, Intravenous infusion of fat emulsion and phosphatides and emulsifying agents, Acta Chir. Scand. Suppl., 278:1 (1961).

10. A. Tokumura, K. Fukuzawa, and H. Tsukatani, Effects of synthetic and natural lysophosphatidic acids on the arterial blood pressure of different animal species, Lipids, 13:572 (1978).

11. J. Y. Wessely, G. Boutet, C. Bordat, Etude comparative des effets hémo dynamiques chez l'animal de deux émulsifiants des lipides utilisés en alimentation parentérale, Gastroentérol. Clin. Biol., 7:205 (1983).

PHOSPHATIDYLCHOLINE: ENDOGENOUS PRECURSOR OF CHOLINE

Steven H. Zeisel

Nutrient Metabolism Laboratory,
Department of Pathology and Pediatrics
Boston University School of Medicine, Boston, MA 02118

Introduction

The choline which eventually reaches brain comes from one of two sources, the diet and pools of choline which have been synthesized *de novo*. As dietary intake can vary, choline must be stored in free and esterified form so as to ensure the maintenance of uninterrupted supplies of choline to tissues. These storage forms of choline, and the choline formed *de novo* are the endogenous sources of choline from which brain derives its supplies of this important amine. Of all the choline esters, phosphatidylcholine (PtdCho) is undoubtedly the largest and most important endogenous storage form of choline (Table 1). In this chapter, I will concentrate on the mechanisms whereby such endogenous pools of phosphatidylcholine serve as precursors of brain choline.

TABLE 1. Estimates of Free Choline and Phosphatidylcholine in the Human.

Organ	Weight (grams)	Choline (nmol/g)	PtdCho (μmol/g)	Choline/organ (μmoles)	PtdCho/organ (mmoles)
Muscle	28,000	107	9.5	2996	267
Adipose	12,500	106	4.8	1325	60
Adrenals	14	344	20.3	5	0.3
Plasma	3,100	10	1.4	31	4
RBC	2,400	50	2.0	120	5
Bone	10,000	-	4.0	-	40
Brain	1,430	40	18.6	57	27
GI tract	1,200	134	3.5	161	4
Heart	330	40	8.5	13	3
Kidney	310	77	6.2	24	2
Liver	1,800	97	18.6	174	33
Lung	1,000	146	2.1	146	2
Other	7,796	50	5.6	390	45
Totals	70 kg			5.4 mmol	492 mmol

In order to estimate the approximate magnitude of pool sizes in the human we used organ weight data from the human (23), choline concentration data from the guinea pig (22), and phosphatidylcholine (PtdCho) concentration data from humans and rats (21).

The Importance of Choline

Choline is a quaternary amine which is necessary for the normal function of the mammalian organism. It is a precursor for the biosynthesis of phospholipids - essential components of all membranes, and it is needed to make acetylcholine, a neurotransmitter (1-3). It is an important constituent of brain, yet more unesterified (free) choline leaves the brain, *in vivo*, than enters it [4-8].

Choline deficiency has major consequences for many species of animals. It is associated with fatty infiltration of the liver in the rat, dog, hamster, pig, baboon and chicken (1,9-12). Fatty acids can be transported out of the liver only when they are properly "packaged" as triglycerides within lipoproteins. An important constituent of these lipoproteins is PtdCho. When adequate supplies of PtdCho are not available the liver is unable to export triglyceride and becomes infiltrated with fat (11,12). The biosynthesis of PtdCho is dependent upon the availability of its precursor, choline. Renal function may also be impaired due to membrane PtdCho deficiency; choline-deficient animals have abnormal concentrating ability, free water reabsorption, glomerular filtration rate, renal plasma flow and gross renal hemorrhaging (13-15). Choline deficiency in animals has also been associated with infertility, growth retardation, bony abnormalities, and increased sensitivity to hepatic carcinogens (1, 10, 13-15). Methyl-donors (such as methionine) can spare some of the choline requirement, as choline can be formed *de novo* by the methyl-ation of phosphatidylethanolamine within liver and other organs (see below) (16,17).

Human cells grown in culture have an absolute requirement for choline (18). No carefully controlled study of the effects of experimentally induced choline deficiency in humans has ever been performed. Malnourished humans have very low choline concentrations in plasma (19). Humans fed with amino acid solutions during total parenteral nutrition take in no choline, and therefore do not rehabilitate their low choline stores (19). In experimental animals, similar changes in plasma choline occur during choline deficiency and are reflected in tissues which can derive choline from blood. Duodenal, cardiac, renal and hepatic levels of choline and choline metabolites are decreased in choline deficient animals (20).   For these reasons we believe that patients being fed parenterally have deficient levels of choline in certain tissues (19, 19a).

Choline as a Pharmacologic agent

Choline is also important because of the effects that supplemental doses of choline have upon organ function. Ingestion of choline increases brain choline concentrations, and influences brain acetylcholine synthesis and release (3,24).   Acetylcholine is the neurotransmitter which mediates such neuronal functions as control of respiration, muscle contraction, heart rate and rhythm, adrenal excretion of epinephrine, gastrointestinal motil-ity, and storage of short term memories. Most neurons which leave the brain (e.g., preganglionic sympathetic, parasympathetic and motor neurons) use this transmitter (1). Choline and PtdCho have been administered in pharmacologic quantities in order to increase cholinergic neurotransmission in humans with neurologic diseases (See chapter by Dr. John Growdon).

Exogenous Sources of Choline

An understanding of dietary sources of choline allows us to place the importance of endogenous pools of choline in perspective. Most mammals derive much of their daily choline requirements from their diets (1), in the form of PtdCho (PtdCho) from foods such as liver, eggs, soybeans and peanuts (1, 25). The adult human ingests 300-1000 mg of choline moiety per day (25, 26). In adults, tissue choline levels are significantly increased following ingestion of a choline or PtdCho-supplemented meal (26). Similarly, neonatal rats in the fed state have higher blood choline levels than those who have been fasted (27). Milk is the first, and often the sole food for the mammalian neonate. It contains large amounts of free choline, PtdCho and sphingomyelin (a choline-containing phospholipid) (62).   Artificial formulas can have a choline content that is very different from that of mother's milk (28). Neonatal animals and humans have exceptionally high blood choline concentration (27, 27a).

Mechanisms exist for transport of choline-containing compounds from gut into blood. In the mature rat, choline is transported into all areas of the small intestine as well into the distal colon (29). At low lumenal choline concentrations, a saturable transport mechanism predominates, while the non-saturable mechanism becomes important at high lumenal choline concentrations. In the colon, we observed only a linear, non-saturable transport process. Our studies show that choline uptake into the jejunum of 10-day old rats occurs in a manner similar to adult transport (i.e., a saturable component active at low choline concentrations, and non-saturable component which predominates at high choline concentrations (29).

<u>Endogenous Sources of Choline originating outside of Brain</u>

Brain can extract choline molecules from the blood. A specific carrier mechanism transports free choline across the blood-brain barrier [30]. Ingestion of choline-containing compounds in the diet results in the elevation of blood and tissue choline levels [31,32]. It is likely that esterified choline also enters brain. Illingworth & Portman observed a small influx of lysoPtdCho from blood to brain [33]. Jope & Jenden [34] observed that, after administration of labelled choline, the specific activity of lysoPtdCho was higher than the specific activity of PtdCho in brain,, strongly suggesting that lysoPtdCho is transported from plasma to brain. However, Pardridge [35] reported that the permeability of the blood-brain barrier to lysoPtdCho or to PtdCho was very low. No arterio-venous difference (across rat brain) was noted for PtdCho or lysoPtdCho [8].

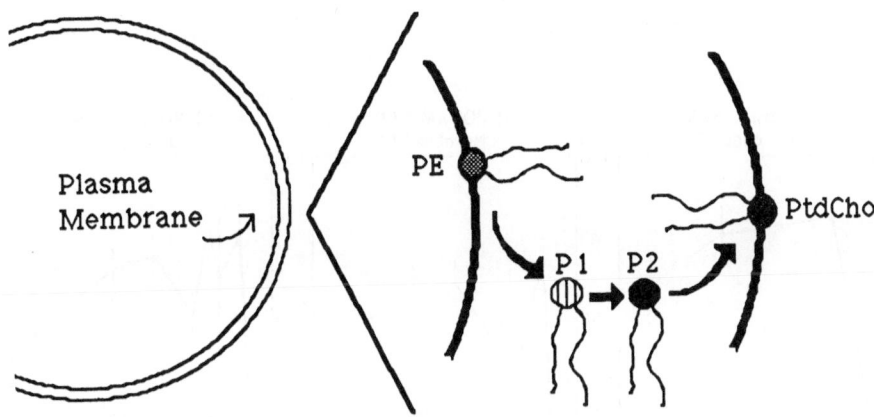

Figure 1: Intermediate Products of PEMT Flip-Flop Across the membrane

Several organs possess the ability to synthesize choline molecules *de novo,* by the sequential methylation of phosphatidylethanolamine, forming PtdCho. Of the 3 enzymatic pathways that catalyze PtdCho biosynthesis, only this one generates new choline molecules. The cytidine diphosphocholine [2] and base exchange [36, 37] pathways do not cause a net synthesis of choline, but only redistribute preexisting molecules. It is only by the methylation pathway, catalyzed by the enzyme(s) phosphatidylethanolamine-N-methyl-transferase (PEMT), that new choline molecules are produced [16, 17]. Almost all (99%) of this enzyme activity is found in the liver [16, 38], although kidney, testes, heart lung, adrenal, erythrocyte, spleen and brain also possess some activity [17, 39]. The liver utilizes the methylation pathway to meet an estimated 15% of the daily requirement for choline in the rat (about 13 µmol/g liver/day). This pathway synthesizes about 15-40% of hepatic PtdCho, the rest being made via the CDP-pathway [40, 41].

PEMT is a complex of at least 2 separate activities, one that adds the first methyl group, and a second activity that adds the final 2 methyl groups. S-adenosylmethionine (AdoMet) acts as the methyl donor for both components [39, 16]. The activity of the first enzyme is limiting [16], and it is located on the inner surface of the plasma membrane leaflet. The second enzyme is located on the outer cell surface, and the phospholipid substrate flip-flops across the membrane as it is methylated [42, 43]. In the liver, PEMT located in the endoplasmic reticulum produces much of the PtdCho exported from the liver. In the endoplasmic reticulum, the first methylation takes place on the inner leaflet, but the enzyme(s) mediating the addition of the last 2 methyl groups are located on both sides of the membrane. [43]

Several factors modulate hepatic PEMT activity, the most important being diet and femaleness (38,40,44). *In vitro* activity of the choline neogenesis pathway is greater during choline deficiency, but this is only seen when exogenous S-adenosylmethionine (AdoMet) is added to the incubation mixture (44). The availability of AdoMet in the liver of choline deficient animals limits the activity of this pathway (45). It is apparent that during choline deficiency AdoMet in liver is depleted, thus markedly limiting the rate of choline neogenesis [12,13,46]. In addition, the activity of this pathway in the liver of newborn rats is very low (see figure 2), while choline requirements for use in growth related membrane formation are very high (27). For this reason, the sequelae of choline deficiency are most easily elicited in young, growing animals (48).

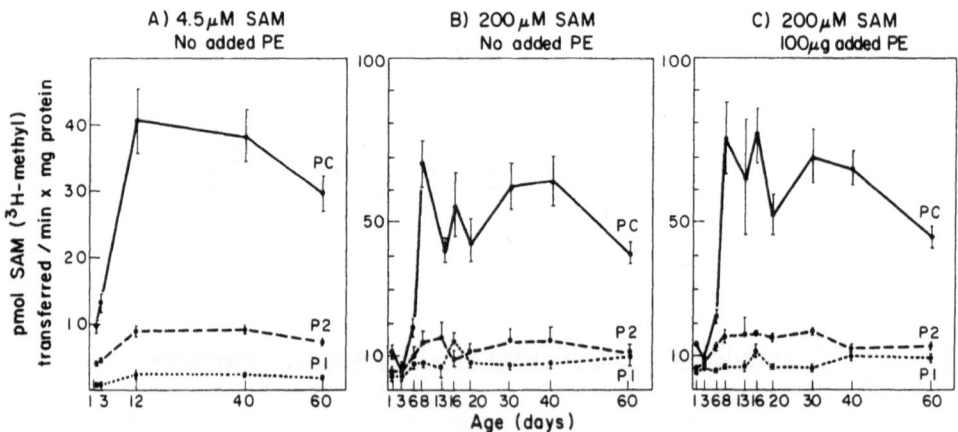

Figure 2: Developmental Changes in Hepatic PEMT
Liver homogenates were prepared from Sprague Dawley rats (both sexes) at the ages noted. PEMT activity was assayed using the method of Hirata (42). In panel A incubations were performed in the presence of endogenous substrates. In panel B S-adenosylmethionine (SAM) was added to acheive a final concentration of 200 µM. In panel C 100 µg/mg protein phosphatidylethanolamine (PE) was added as well as S-adenosylmethionine (SAM) to acheive a final concentration of 200 µM. All incubations were for 15 min at 37°C. 3 rats were used per point. Data are expressed as mean ± SEM.

Release of Choline from PtdCho in the Liver

In the liver the major pathway for metabolism of PtdCho is via phospholipase A1 (EC3.1.1.32), and A2 (EC 3.1.1.4) and lysophospholipase (EC 3.1.1.5). The resultant glycerophosphorylcholine is hydrolyzed by a diesterase (EC 3.1.4.38) (49). Liver exports phospholipids in 2 secretions: plasma lipoproteins, and bile. The amount of PtdCho exported per day in humans is approximately 10-20% of the liver PtdCho pool, divided equally between the two secretions (50). Because the PtdCho excreted in bile leaves the body (enters the lumen of the gut), we will not consider it as an endogenous source for choline. The liver packages triglyceride for export in a lipoprotein envelope rich in PtdCho (very low density lipoprotein, VLDL). Circulating high density lipoprotein (HDL) transfers apoprotein C to VLDL, this apoprotein activates lipoprotein lipases which are in capillary endothelia. This results in the hydrolysis of triglyceride, so that the ratio of envelope (and thus PtdCho) to triglyceride increases. Thus low density lipoprotein (LDL) and HDL lipoproteins are rich in PtdCho content. Plasma contains aproximately 1.5 mM PtdCho (unpublished observations). Some of the PtdCho released by liver into the blood is metabolized to form lysolecithin by phospholipase A2 activity in plasma, and by lecithin-cholesterol acyl transferase which has been released by the liver (51). LysoPtdCho is bound to albumin within blood, and is removed extremely rapidly from blood, but it is possible that small amounts survive long enough to enter brain (51).

Endogenous Sources of Choline in Brain

More unesterified (free) choline leaves the brain, *in vivo*, than enters it; there is a net efflux of 7 nmol/g/min [4-8, 34, 52]. Homogenates, slices, and other *in vitro* preparations of brain produce similar amounts of choline when incubated at 37°C [7,8, 34,52]. Some of this choline may come from choline-esters in blood, but most must come from pools of choline within brain and from *de novo* synthesis.

The total pool of free choline within normal rat brain (25 nmol/g wet wt) would be quickly exhausted if it were the sole source of choline efflux [53]. The same is true for acetylcholine (20-70 nmol/g wet wt), CDP-choline (50 nmol/g wet wt), phosphocholine (380 nmol/g wet wt), glycerophosphocholine (400 nmol/g wet wt), and choline plasma-logen (600 nmol/g wet wt) [4,53]. Membranes, which contain sphingomyelin (3700 nmol/g wet wt), and PtdCho (PtdCho; 15,000 nmol/g wet wt), might be reservoirs from which free choline could be derived [4]. If this is true, then choline accumulation occurs at the expense of membrane breakdown.

PEMT activity is present in membranous fractions of rat and bovine brain (54), again reflecting at least two enzyme activities, distinguished by their substrate and kinetic properties. The first enzyme, located on the cytoplasmic side of the synaptosomal membrane catalyzes the addition of the firts methyl group, whereas the active site of the second and third methylation is on the outside of the synaptosome membrane (54). The first methyltransferase has a high affinity for AdoMet (2-4 µM), wheras the second enzyme has a low affinity for the cofactor (Km= 20-110 µM) (54). The enzyme catalyzing the first methylation is the rate limiting step in PtdCho synthesis in rat brain (55).

The orientation of the PEMT system is such that choline is delivered outside of the plasma membrane. There is good evidence that acetylcholine synthesis within the nerve terminal is dependent upon choline supplied externally to the plasma membrane. If the uptake of choline into the terminal is blocked with HC-3, acetylcholine synthesis ceases (56).

In neonatal brain the relationships between AdoMet concentrations and PEMT activity (first methyaltion) are more complex (Figure 3) (55). A sigmoidal component (hill coefficient of 2.7), requiring 90µM AdoMet for half saturation, predominated over the high affinity component (similar to that of adult brain). This unusual form of PEMT could not be detected in brains of rats older than 5 days of age. This special form of PEMT is present in large amounts, and the rate of PtdCho synthesis via the PEMT pathway (given that AdoMet concentrations in neonatal brain are 40-50 nmol/g) is highest during the first days of life. Thus neonates, which need a lot of choline for brain growth, possess a special form of PEMT to facilitate choline synthesis.

The activity of PEMT may be regulated by various neurotransmitters. When synaptosomes were incubated with dopamine (100 µM), the methylation of phosphatidylethanolamine increased by almost 2 fold.   The effect of dopamine was observed at concentrations as low as 1 μM, and could be blocked by haloperidol (57). There is no sexual dimorphism in PEMT activity in brain (55).

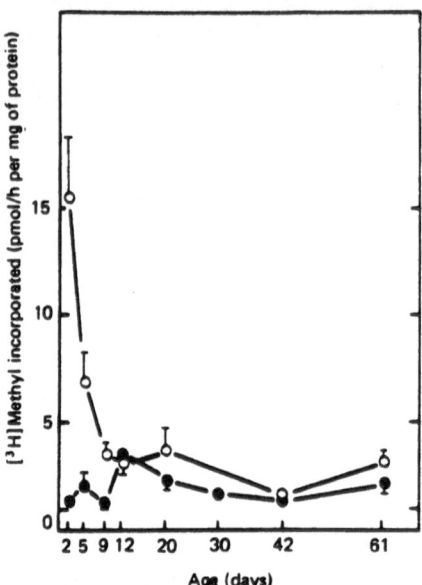

Figure 3: Developmental Changes in Activity of PEMT in Rat Brain.
Brain homogenates were incubated in the presence of 3.3 μM (filled circles) or 200 μM (open circles) [methyl-3H]-AdoMet. Methylated intermediates were isolated by thin layer chromatography. Data are expressed as pmol/h per mg protein ± SEM. Taken from reference 55.

Net efflux of choline from brain is substantial, and the methyltransferase activity in brain is relatively small. Assuming that the methyltransferases are operating at maximal velocity *in vivo* , about 10% of choline efflux from brain to plasma could be accounted for by PEMT activity (54, 58).

Enzymatic mechanisms for Releasing the Choline Moiety of PtdCho

Even though estimates of the relative contribution of peripheral versus brain sources of choline molecules are a matter of controversy, the only way that there could be a net efflux of choline from brain is if mechanisms exist within brain for the hydrolysis of choline-esters.

Estimates of the turnover rate of the PtdCho in synaptosomes indicate that there are 2 pools in nerve endings: one with a half life of 2 days the other with a half-life of 53 days (59)   Such measurements of turnover do not take into account the recycling of parts of PtdCho during this phospholipid's synthesis. Synaptosomes incubated with radiolabelled AdoMet and phosphatidyldimethylethanolamine make PtdCho, and then rapidly cleave it to form unesterified choline (58). Free choline constitutes 23% of labelled choline formed in this system after less than an hour of incubation. This suggests that the turnover rate of certain pools of PtdCho can be exceedingly rapid.

Rat brain is capable of producing large amounts of free choline, and that ATP concentrations within brain modulate the accumulation of choline (52). Many studies have demonstrated choline accumulation within preparations of brain, heart, ileum, kidney, lung, and liver during *in vitro* incubations [31,34, 52, 60-65]. In brain and heart, choline formation is stimulated by activation of muscarinic acetylcholine receptors [63, 65], but is not related to the degree of cholinergic innervation of the tissue [61].

In brain, free choline formation is mediated by enzymatic activities, as treatments known to inactivate enzymes (microwave irradiation, heat) stop the postmortem accumu - lation of choline (52). Choline formation is temperature and pH dependent (52).

Choline liberation within brain reflects the contributions of several parallel pathways, including the breakdown of glycerophosporylcholine, acetylcholine and PtdCho. PtdCho, and not other choline-esters, is the most important reservoir from which free choline is produced, though glycerophosphorylcholine can also be a precursor (4, 34, 52). The pool of PtdCho present in brain homogenates is 40 times larger than the amount of choline produced during an hour of incubation at 37°C (52). Of all the choline-containing metabolites in brain only PtdCho pool size decreases enough (15-20%) to account for choline formation (4, 34, 52).

There are many possible routes for PtdCho metabolism which could result in the formation of free choline. There are reasons to believe that phospholipase A2 and phospholipase D activities are important for the liberation of choline from PtdCho in brain. Phospholipase D activity (EC 3.1.4.4) forms free choline and phosphatidic acid from PtdCho (66,67). Activity is found in brain microsomes [33] and is most enriched in synaptosomal membranes (66, 67). Activation of this enzyme requires the presence of added detergent or fatty acids, especially oleic acid [6,67].

PtdCho can be deacylated in the 2-position by phospholipase A2 (EC 3.1.1.4), an enzyme activity associated with many membranes and which is stimulated by, but does not require, calcium and which is inhibited by zinc [68-71]. The activity of phospholipase A2 is increased when various receptors on plasma membranes are activated (70). Phospholipase A1 activity, which deacylates PtdCho in the 1-position, is widely distributed in brain. It is most active at acid pH, and does not require calcium [71]. It is likely that phospholipase A activity is the primary initiator of choline release from PtdCho. The stimulation of phospholipase A2, by releasing fatty acids, may activate phospholipase D and thereby accelerate the liberation of choline from PtdCho.

Phospholipase C activity (EC 3.1.4.3), which forms phosphorylcholine and diacylglycerol, requires, and is stimulated by calcium. Maximal activity is noted at pH 8, and it is a cytosolic enzyme [72]. In brain, the cytosolic fraction of brain is not required for choline formation, suggesting that this enzyme was not the predominant activity responsible for free choline generation (52). PtdCho can be metabolized by reversal of the "base"-exchange pathway, in which bound choline is liberated and replaced with another moiety, such as serine. The pH optimum of this pathway is 9.0, and calcium is required [73]. We noted no increase in choline formation when serine and calcium were added to incubations of brain homogenates (52). PtdCho can also be degraded by reversal of the activity of diacylglycerol:CDP-choline phosphotransferase (EC 2.7.8.2) whenever large amounts of CMP are present [74]. We duplicated the conditions necessary for optimal activity of this pathway (added CMP, magnesium and dithiothreitol) and we noted no increase in formation of free choline by brain homogenates.

As discussed earlier, the pools of acetylcholine and CDP-choline within brain are too small to be the sole source of free choline [52]. Phosphorylcholine can be degraded to release free choline in reactions catalyzed by alkaline phosphatase (EC 3.1.3.1) and acid phosphatase (EC 3.1.3.2) [75]. We demonstrated a significant excess capacity for the liberation of choline from phosphorylcholine within brain (52). However, though phosphorylcholine might be an immediate source of choline, it is unlikely that this was the ultimate pool from which choline molecules were derived, as we (52), and others [34] have observed a net increase in phosphorylcholine concentrations during incubations.

Brain glycerophosphorylcholine can be degraded to form choline and glycero-3-phosphate, catalyzed by glycerophosphorylcholine-diesterase (EC 3.1.4.2) [76,77]. During incubation of brain tissue, there is a significant, but transient fall in glycerophosphorylcholine concentration (4, 34, 52). There are several reasons to believe that glycerophosphorylcholine is not the only reservoir from which choline can be derived. The fall in glycerophosphorylcholine content would only account for a small fraction of the free choline formed by brain. Very low concentrations of magnesium (less than 18 μM) [4] are required for activity of glycerophosphorylcholine-diesterase. Larger amounts of magnesium should not further increase diesterase activity [77], yet we noted significant stimulation of choline production by magnesium, with saturating concentrations being 3-5 mM (52). Zinc ions do not inhibit glycerophosphorylcholine-diesterase activity [77], yet we found that zinc was a potent inhibitor of choline formation, even at concentrations as low as 100 μM (52). Glycerophosphorylcholine is a water soluble compound, and should not be found in particulate subcellular fractions of brain. We found that all membrane-containing subcellular fractions of brain produced free choline at approximately the same rate, and that there were no cofactors or enzyme activities present in the soluble fraction that increased choline production (52). This findings make it unlikely that glycerophosphorylcholine was the ultimate or sole source from which choline was formed.

LysoPtdCho is a membrane-toxic compound, and is never present in large amounts in brain [33,52]. It is hydrolyzed to release free choline 8-20 times faster than it is acylated to form PtdCho (78). We never observed accumulation of significant quantities of lysoPtdCho and could not detect its formation when radiolabeled PtdCho was added to incubation solutions (52).

Sphingomyelin is present in brain, and it can be degraded by sphingomyelinase (EC 3.1.4.1.2). The non-lysosomal form of this enzyme is magnesium dependent and is activated by Triton x100 [79]. No activity has been reported in brain homogenates in the absence of detergent [80]. We observed formation of free choline from radiolabeled sphingomyelin (0.3 nmol/mg protein/hr, however, this did not result in detectable reductions in brain sphingomyelin concentrations (52).

ATP is an inhibitor of choline formation from endogenous sources, and from added PtdCho, lysoPtdCho, phosphorylcholine and glcerophosphorylcholine (52). There are several possible mechanisms whereby ATP could inhibit the accumulation of choline. It might chelate magnesium, thereby making this trace metal unavailable to an enzyme which required it. Conversely, magnesium may have activated an ATPase, thereby decreasing concentrations of ATP. Under steady-state conditions the net activities of the "forward" enzymes of PtdCho synthesis (choline kinase, phosphocholine cytidyltransferase, diacylglycerol:CDP-choline phosphotransferase, and phosphatidylethanolamine-N-methyltransferase) must equal the net activities of the "reverse" enzymes (phospholipases A1, A2, C and D, glycerophosphocholine-diesterase, lysophospholipase and lysophospholipase D). Net production of free choline could occur if utilization of choline for the synthesis of esters was inhibited. ATP is required for the synthesis of phosphorylcholine and for the synthesis of CDP-choline, two intermediates in the Kennedy pathway of PtdCho synthesis. The addition of ATP to brain homogenates did increase choline-ester synthesis, but not enough to account for the degree of inhibition of free choline accumulation that was observed (52). For this reason we conclude that ATP inhibited choline-ester catabolism as well as enhanced choline-ester synthesis. Probably, in the presence of adequate ATP concentrations, most lysoPtdCho was reacylated, making little available for further degradation to form free choline. Reacylation requires ATP to form the acyl-CoA moiety, and is rapid (2 nmol/mg protein/min of PtdCho formed) [81]. Lysophospholipase D, an enzyme which forms choline from lysoPtdCho, is inhibited by zinc and is stimulated by magnesium [52].

We have suggested that ATP may be an important modulator of PtdCho metabolism. If this is the case, then other parts of the PtdCho molecule, such as free fatty acids, should accumulate after treatments which lower ATP availability. There is a large increase in free fatty acid concentrations in brain within 30 seconds after interruption of the supply of arterial blood to brain [82].

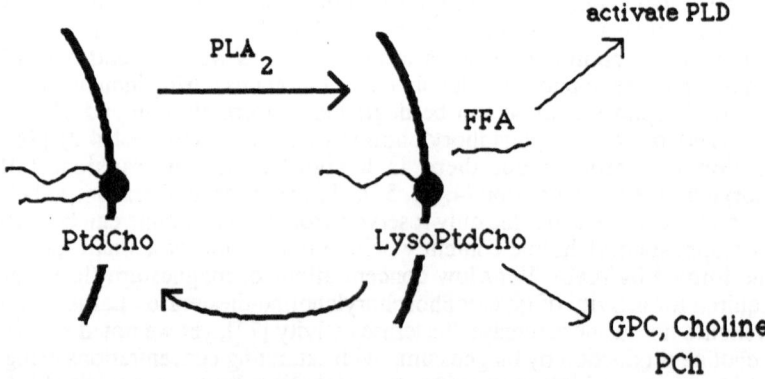

Figure 4: Probable Cleavage of PtdCho by Phospholipases A2 and D

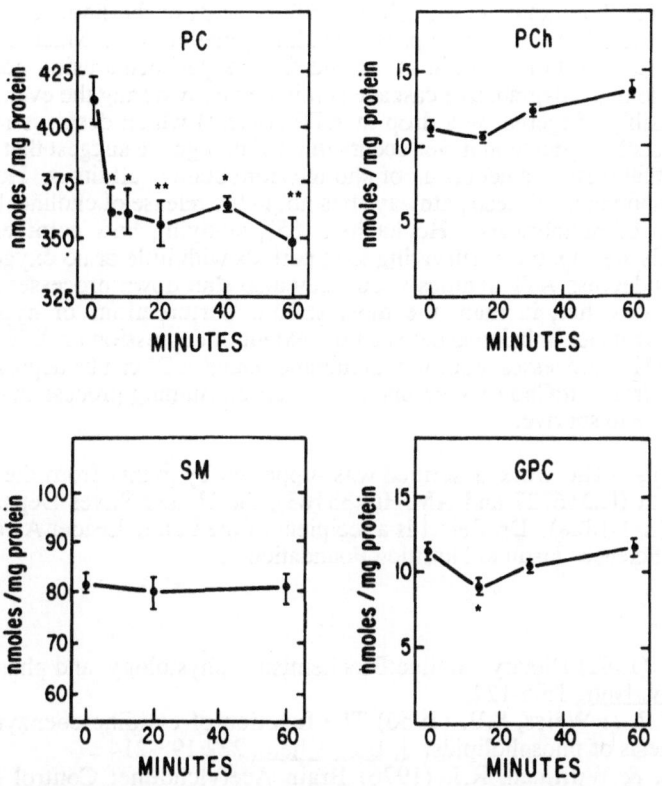

Figure 5: Changes in Choline-Containing Compounds in Brain During Incubation at 37 °C.

Homogenates of brain were incubated at 37 °C for the time indicated. Choline containing compounds were isolated using ion exchange chromatography. They were hydrolyzed to release choline, which was assayed using a radio-enzymatic assay. Data are expressed as mean ± SEM. Data from reference 52.

Summary

        We have described how brain derives choline from endogenous PtdCho stores. These mechanisms are of importance because choline plays such a vital role in membrane and acetylcholine synthesis within brain. It is possible that the release of choline from PtdCho is carefully regulated so as to be linked with choline-requirements for acetylcholine synthesis. Brain slices deprived of choline, but stimulated to release acetylcholine, cannibalize their membrane PtdCho (83). It is more likely that the catabolism of PtdCho is triggered by many factors, some of which may be independent of the need for choline. For example, the release of fatty acids of the linoleic series from PtdCho may be the rate limiting step in prostaglandin formation (84). Much of what we know about choline liberation in brain is derived from experiments using postmortem tissue. It is possible that a degradative cascade is initiated early during the events leading to cell death (possibly triggered by a drop in ATP content) which destroys membrane PtdCho. Perhaps such degradation never occurs *in vivo,* though we suggest that it must if we are to believe that there is a net efflux of choline from brain at all times. However, it may be that the importance of these pathways lies not in the release of choline, but rather in the destruction of membranes. Hochachka (85), studying how ectothermic and hibernating animals are capable of surviving long periods with little or no oxygen, found that these animals depress ATP synthesis, but they also shut down processes requiring ATP. He speculates that, in man, the most serious perturbations of hypoxia and hypothermia arise from an imbalance between the extent of depression of ATP synthesis and the extent to which processes requiring membrane-bound ATP can be depressed. The release of choline from PtdCho may be one such ATP-consuming process that must be supressed if cells are to survive.

Acknowledgements   The work described was supported by grants from the National Institutes of Health (HD16727 and AM/HD 33163), the United States Department of Agriculture (CRCR-1-1828). Dr. Zeisel is a recipient of the Future Leader Award of the International Life Sciences Institute-Nutrition Foundation.

References

1.   Zeisel, S.H. (1981) Dietary Choline:Biochemistry, physiology, and pharmacology. Ann. Rev. Nutr. 1:95-121.
2.   Kennedy, E.P. & Weiss, S.B. (1956) The function of cytidine coenzymes in the biosynthesis of phospholipids. J. Biol. Chem. 222:193-214 .
3.   Cohen, E.L. & Wurtman, R.J. (1976) Brain Acetylcholine: Control by dietary choline. Science 19:561-562 .
4.   Ansell, G.B. and Spanner, S. Choline transport and metabolism in brain, in Phospholipids in the Nervous System (Horrocks, L.A., Ansell, G.B., and Porcellati, G., eds.), Vol.1, pp. 137-144, Raven, N.Y. (1982).
5.   Dross, K., Kewitz, H.(1972) Concentration and origin of choline in rat brain. N.S. Arch. Pharmacol. 274, 91-106.
6.   Choi, R.L., Freeman, J.J. and Jenden, D.J. (1975) Kinetics of plasma choline in relation to turnover of brain choline and formation of acetylcholine. J. Neurochem. 24, 735-741.
7.   Aquilonius, S.M., Ceder, G., Lying-Tunnell, U., Malmud, H.O., Shubert, J. (1975) The arteriovenous difference of choline across the brain of man. Brain Res. 99, 430-433.
8.   Spanner, S., Hall, R. Ansell, G.B. (1976) Arterio-venous differences of choline and choline lipids across the brain of rat and rabbit. Biochem. J. 154:133-140.
9.   Atsushi, I., Hellerstein, E.E., Hegsted, D.M. (1963) Composition of dietary fat and the accumulation of liver lipid in the choline-deficient rat. J. Nutr. 79:488-492.
10. Best, C.H. & Huntsman, M.E. (1932) The effects of the components of lecithin upon the deposition of fat in the liver. J. Physiol. 75:405-412.
11. Lombardi, B. (1971) Effects of choline deficiency on rat hepatocytes. Fed. Proc. 30:139-142.
12. Haines, M. (1966) The effects of choline deficiency and choline re-feeding upon the metabolism of plasma and liver lipids. Can. J. Biochem. 44:45-57.

13. Michael, U.F., Cookson, S.L., Chavez, R., Pardo, V. (1975) Renal function in the choline deficient rat. Proc. Soc. Exp. Biol. Med. 150:672-676 .

14. Baxter, J.H. (1947) A study of hemorrhagic-kidney syndrome of choline deficiency. J. Nutr. 34:333.

15. Best, C.H., Hartroft, W.S. (1949). Symposium on nutrition in preventative medicine: Nutrition, renal lesions and hypertension. Fed. Proc. 8:610.

16. Bremer, J., Greenberg, D. (1961) Methyltransferring enzyme system of microsomes in the biosynthesis of lecithin. Biochim. Biophys. Acta 46:205-216.

17. Blusztajn, J.K., Zeisel, S.H., Wurtman, R.J. (1979) Synthesis of lecithin (phosphatidylcholine) from phosphatidylethanolamine in bovine brain. Brain Res. 179:319-327.

18. Eagle, H. (1955) The minimum vitamin requirements of the L and Hela cells in tissue culture, the production of specific vitamin deficiencies, and their cure. J. Exp. Med. 102:595-600 .

19. Sheard, N.F., Tayek, J.A., Bistrian, B.R., Blackburn, G.L., Zeisel, S.H. (1985) Plasma choline concentrations in humans fed parenterally. Am. J. Clin. Nutr. 42:352-60.

19a. Burt, M.E., Hanin, I., Brennan, M.F. (1980) Choline deficiency associated with total parenteral nutrition. Lancet ii:638-639.

20. Haubrich, D.R., Wang, P.F., Clody, D.E. (1975) Increase in rat brain acetylcholine induced by choline or deanol. Life Sci. 17:975-980.

21. White DA. (1973) The phospholipid content of mammalian tissues, in Form and Function of Phospholipids (Ansell, G.B., Hawthorne, J.N., and Dawson, R.M.C., eds.), pp. 441-482, Elsevier, Amsterdam.

22. Haubrich, D.R., Wang, P.F.L., Wedeking, P.W. (1975) Distribution and metabolism of intravenously administered choline-[methyl-3H] and synthesis of acetylcholine in various tissues of guinea pigs. J. Pharm. Exp. Ther. 193:246-255.

23. Report on Task Force on Reference Man (Snyder, et al. eds.) Pergamon, Oxford, (1975).

24. Haubrich, DR, Wang, PF, Chippendale, T. Procter, E. (1976) Choline and acetylcholine in rats: Effect of dietary choline. J. Neurochem. 27:1305-1313.

25. Wurtman, J.J. (1979) Sources of choline and lecithin in the diet, in Nutrition and the Brain, Vol. 5 (Wurtman & Wurtman, eds.), Raven, N.Y. pps. 73-81.

26. Zeisel, S.H., Growdon, J.H. Wurtman, R.J., Magil, S.G., Logue, M. (1980) Normal plasma choline responses to ingested lecithin. Neurology 30:1226-1229.

27. Zeisel, S.H., Wurtman,R.J. (1981) Developmental changes in rat blood choline concentration. Biochem. J. 198:565-570.

27a. Mallinger, A.G., Hanin, I., Stumf, R.L., Mallinger, J., Kopp, U., Erstling, O. (1983) Lithium treatment during pregnancy: a case study of erythrocyte choline content and lithium transport. J. Clin. Psychiatry 44:381-384.

28. Zeisel, S.H., Char, D., Sheard, N.F. (1986) Choline, phosphatidylcholine and sphingomyelin in human and bovine milk and infant formulas. J. Nutr. 116:50-58.

29. Sheard, N.F., Zeisel, S.H. (1986) An in vitro study of choline uptake by intestine from neonatal and adult rats. Ped. Res. 20:768-772.

30. Cornford, E.M., Braun, L.D., Oldendorf, W.H. (1978) Carrier mediated transport of choline and certain choline analogs. J. Neurochem. 30:299-308.

31. Haubrich, DR, Wang, PF, Chippendale, T. Procter, E. (1976) Choline and acetylcholine in rats: Effect of dietary choline. J. Neurochem. 27:1305-1313.

32. Haubrich, D.R., Gerber, N., Pflueger, A.B., Zweig, M. (1981) Tissue choline studied using a simple chemical assay. J. Neurochem. 36:1409-1417.

33. Illingworth, D.R., Portman, O.W. (1973) Formation of choline from phospholipid precursors: A comparison of the enzymes involved in phospholipid catabolism in the brain of the rhesus monkey. Physiol. Chem. Physics 5:365-373.

34. Jope, R.S., Jenden, D.J. (1979) Choline and phospholipid metabolism and the synthesis of acetylcholine in rat brain. J. Neurosci. Res. 4, 69-82.

35. Pardridge, W.M., Cornford, E.M., Braun, L.D., Oldendorf, W. (1979) Transport of choline and choline analogues through the blood-brain barrier, in Nutrition and

the Brain, vol 5. (A. Barbeau, J.Growdon, R. Wurtman, eds.), Raven Press p.25-34.

36. Orlando P., Arienti G., Cerrito, F., Massari P., Porcellati G. (1977) Quantitative evaluation of two pathways for phosphatidylcholine biosynthesis in rat brain in vivo. Neurochem. Res. 2:191-201.

37. Salerno D.M., Beeler D.A. (1973) The biosynthesis of phospholipids and their precursors in rat liver involving de novo methylation and base exchange pathways in vivo. Biochim. Biophys. Acta 326:325-338.

38. Bjornstad P., Bremer J. (1966) In vivo studies on pathways for the biosynthesis of lecithin in the rat. J. Lipid Res. 7:38-45.

39. Hirata F., Axelrod, J. (1980) Phospholipid methylation and biological signal transmission. Science 209:1082-1090.

40. Linblad L, Schersten T. (1976) Incorporation rate in vitro of choline and methyl-methionine into human hepatic lecithin. Scand. J. Gastroenterol. 11:587-591.

41. Sundler R, Akesson B. (1975) Regulation of phospholipid synthesis in isolates of rat hepatocytes. J. Biol. Chem. 250:3359-3367.

42. Hirata F., Axelrod, J. (1978) Enzymatic synthesis and rapid translocation of phosphatidylcholine by two methyltransferases in erythrocyte membranes. Proc. Natl. Acad. Sci. USA 75:2348-2352.

43. Higgins, J.A . (1981) Biogenesis of endoplasmic reticulum phosphatidylcholine. Translocation of intermediates across the membrane bilayer during methylation of phosphatidylethanolamine. Biochim. Biophys. Acta 640:1-15.

44. Schneider, WJ, Vance, DE. (1978) Effect of choline deficiency on the enzymes that synthesize phosphatidylcholine and phosphatidylethanolamine in rat liver. Europ. J. Biochem. 85:181-187.

45. Pascale, R., Pirisi, L., Daino, L., Zanetti, S., Satta, A., Bartoli, E., Feo, F. (1982) Role of phosphatidylethanolamine methylation in the synthesis of phosphatidylcholine by hepatocytes isolated from choline-deficient rats. FEBS Letters 145:293-297.

46. Lyman, R.L., Sheenan, G., Tinoco, J. (1973) Phosphatidylethanolamine metabolism in rats fed a low methionine, choline-deficient diet. Lipids 8:71-79.

47. Cornatzer, W.E., Hoffman, D.R., Haning, J.A. (1984) The effect of hyper and hypothyroidism, hypophysectomy and adrenalectomy on phosphatidylethanolamine methyltransferase, phosphatidylmethyl-ethanolamine methyltransferase and choline Lipids 19:1-4.

48. Griffith, W.H., Dyer, H.M. (1968) Present knowledge of methyl groups in nutrition. Nutr. Rev. 26:1-8.

49. Dawson, R.M.C. (1955) Role of glycerolphosphorylcholine and glyceryl-phosphorylethanolamine in liver phospholipid metabolism. Biochem J. 59:5-8.

50. Coleman, R. (1973) Phospholipids and the hepato-portal system, in Form and Function of Phospholipids (Ansell, G.B., Hawthorne, J.N., and Dawson, R.M.C., eds.), pp. 345-376, Elsevier, Amsterdam.

51. Stein, Y., Stein, O. (1966) Metabolism of labeled lysolecithin, lysophosphatidyl ethanolamine and lecithin in the rat. Biochim. Biophys. Acta 116:95-107.

52. Zeisel, S.H. (1985) Formation of unesterified choline by rat brain. Biochim. Biophys. Acta 835:331-343.

53. Ansell, G.B. (1973) Phospholipids and the nervous system, in Form and Function of Phospholipids (Ansell, G.B., Hawthorne, J.N., and Dawson, R.M.C., eds.), pp. 377-422, Elsevier, Amsterdam.

54. Blusztajn, J.K., Wurtman, R.J. (1983) Choline and cholinergic neurons. Science 221:614-620.

55. Blusztajn J.K., Zeisel S.H., Wurtman, R.J. (1985) Developmental changes in the activity of phosphatidylethanolamine N-methyltransferases in rat brain. Biochem. J. 232:505-511.

56. Guyenet, P., Lefresne, P., Rossier, J., Beaujouan, J.C., Glowinski, J. (1973) Effect of sodium, hemicholinium-3 and antiparkinson drugs on [14C]-acetylcholine synthesis and [3H]-choline uptake in rat striatal synaptosomes. Brain Res 62:523-529.

57. Leprohan, C.E., Blusztajn, J.K., Wurtman, R.J. (1983) Dopamine stimulation of phosphatidylcholine (lecithin) biosynthesis in rat brain neurons. Proc. Natl. Acad. Sci. USA 80:2063-2065.

58. Blusztajn, J.K., Wurtman, R.J. (1981) Choline biosynthesis by a preparation enriched in synaptosomes from rat brain. Nature 290, 417-418.

59. Pasquini, J.M., Krawiec, E.F., Soto, J. (1973) Turnover of phosphatidylcholine in cell membranes of adult rat brain. J. Neurochem 21:647-652.

60. Browning, E.T. (1971) Free choline formation by cerebral cortical slices from rat brain. Biochim. Biophys. Res. Commun. 45, 1586-1590.

61. Kosh, J.W., Dick, R.M. and Freeman, J.J. (1980) Choline post-mortem increase: Effect of tissue, agitation, pH and temperature. Life Sciences 27, 1953-1959.

62. Collier, B., Poon, P. and Salehmoghaddam, S. (1972) The formation of choline and of acetylcholine by brain in vitro. J. Neurochem. 19:51-60.

63. Dolezal, V., Tucek, S. (1984) Activation of muscarinic receptors stimulates the release of choline from brain slices. Biochem. Biophys. Res. Comm.120:1002-1007.

64. Bhatnager, S.P., MacIntosh, F.C. (1967) Effects of quaternary bases and inorganic cations on acetylcholine synthesis in nervous tissue. Can. J. Physiol. Pharmacol. 45:249-268.

65. Corradetti, R., Lindmar, R., and Loffelholz, K. (1983) Mobilization of Cellular Choline by Stimulation of Muscarine Receptors in Isolated Chicken Heart and Rat Cortex In Vivo J. Pharmacol. Exptl. Therap. 226:826-832.

66. Hattori, H., Kanfer, J.N. (1984) Synaptosomal phospholipase D: potential role in providing choline for acetylcholine synthesis. Biochem. Biophys Res Comm. 124:945-949.

67. Hattori, H., Kanfer, J.N. (1985) Synaptosomal phospholipase D potential role in providing choline for acetylcholine synthesis. J. Neurochem. 45:1578-1584.

68. De Haas, G.H., Postema, N.M., Nieuwenhuizen, W., Van Deenen, L.L.M. (1968) Purification and properties of phospholipase A from porcine pancreas. Biochim. Biophys. Acta 159:103-117.

69. Zahler, P., Kramer, R. (1981) Isolation of phospholipase A2 from red cell membranes of sheep. Meth. Enzym. 71:690-698.

70. Van den Bosch, H. (1980) Intracellular Phospholipases A. Biochim. Biophys. Acta 604, 191-246.

71. Cooper, M.F. and Webster, G.R. (1970) The differentiation of phospholipase A1 and A2 in rat and human nervous tissues. J. Neurochem. 17:1543-1554.

72. Edgar, A.D. and Freysz, L. (1982) Phospholipase activities of rat brain cytosol. Occurrence of phospholipase C activity with phosphatidylcholine. Biochim. Biophys Acta 711, 224-228.

73. Kanfer, J.N. (1982) The base exchange enzymes and phospholipase D of rat brain, in Phospholipids in the Nervous System (Horrocks, L.A., Ansell, G.B., and Porcellati, G., eds.), Vol. 1, pp. 13-20, Raven, N.Y.

74. Goracci, G., Francescangeli, E., Horrocks, L.A., Porcellati,G. (1981) The reverse reaction of choline phosphotransferase in rat brain microsomes. A new pathway for degradation of phosphatidylcholine. Biochim. Biophys. Acta 664:373-379.

75. McFarlane, M.G., Petterson, L.M.B., Robison, R. (1934) The phosphatase activity of animal tissues. Biochem. J. 28:720-724.

76. Baldwin, J.J., Cornatzer, W.E. (1968) Rat kidney glycerophosphorylcholine diesterase. Biochim. Biophys. Acta 164:193-195.

77. Webster, G.R., Marples, E.A, Thompson, R.H.S. (1957) Glycerophosphorylcholine diesterase activity of nervous tissues. Biochem. J. 65:374-377.

78. Portman, O.W., Illingworth D.R., Alexander, M. (1973) Lysolecithin and sphingosinephosphorylcholine in the metabolism of brain phospholipids of the rhesus monkey (Macaca mulatta): Effects of development. J. Neurochem. 20:1659-1667.

79. Gatt, S. (1982) Studies on sphingomyelinase, in Phospholipids in the Nervous System (Horrocks, L.A., Ansell, G.B., and Porcellati, G., eds.), Vol. 1, pp. 181-198, Raven, N.Y.

80. Abra, R.M. and Quinn, P.J. (1975) A novel pathway for phosphatidylcholine catabolism in rat brain homogenates. Biochim. Biophys. Acta 380, 436-441.

81. Fisher, S.K., Doherty, F.J., Rowe, C.E. (1982) Deacylation and acylation of phospholipids in the nervous system, in Phospholipids in the Nervous System (Horrocks, L.A., Ansell, G.B., and Porcellati, G., eds.), Vol. 1, pp. 63-74, Raven, N.Y.

82. Bazan, N.G., Bazan, H.E.P., Kennedy, W.G., Joel, C.D. (1971) Regional distribution and rate of production of free fatty acids in rat brain. J. Neurochem. 18, 1387-1393.

83. Maire, J.C., Wurtman, R.J., (1985) Effects of electrical stimulation and choline availability on the release and contents of acetylcholine and choline in superfused slices from rat striatum. J. Physiol (Paris) 80:189-195.

84. Vogt, W. (1978) Role of phospholipase A2 in prostaglandin formation. In: Advances in Prostaglandin and Thromboxane Research, C. Galli ed., pp. 88-95 Raven press, NY.

85. Hochachka, P.W. (1986) Defense strategies against hypoxia and hypothermia. Science 231:234-241.

USE OF PHOSPHATIDYLCHOLINE IN BRAIN DISEASES: AN OVERVIEW

John H. Growdon

Massachusetts General Hospital
Department of Neurology
Boston, Massachusetts 02114

Two biochemical effects of phosphatidylcholine (PC) account for its
use in clinical trials treating brain diseases: PC as a precursor for
acetylcholine (ACh) biosynthesis, and PC as an integral part of neuronal
membranes. This chapter reviews the scientific basis for each postulated
use, and describes the results of PC administration to patients with
neurological and psychiatric diseases.

PC AS A PRECURSOR FOR ACh BIOSYNTHESIS

The rationale for administering PC to patients with neurological and
psychiatric disorders is based upon the observation that pharmacological
amounts of PC provide an exogenous source of choline that enhances
biosynthesis of the neurotransmitter ACh. ACh is synthesized from a
molecule of choline combining with acetyl coenzyme A in a reaction
catalyzed by the enzyme choline acetyltransferase. This reaction also
requires adequate supplies of oxygen and glucose; if either is restricted,
ACh synthesis is reduced (42). Under physiological conditions, however,
the major rate limiting commodity is choline, and choline is the only one
of the three substances whose administration increases brain ACh levels.
Cohen and Wurtman (25,26) first showed that choline administered by
intraperitoneal injections or added to the diet increased blood choline,
brain choline, and brain acetylcholine levels in rats. Although elevations
in ACh levels occur in whole brain (25,55) and hippocampus (59), the
effects of choline administration are most striking in the striatum.
Wecker et al. (119) reported that choline administration reversed the
decrease in acetylcholine content within striatum caused by anticholinergic
drugs and London et al. (84) found that choline administration increased
ACh levels after neurotoxin-induced striatal neuronal degeneration. PC is
the naturally-occurring source of dietary choline; its administration also
increases brain ACh levels (60). The evidence relating PC (and choline) to
acetylcholine biosynthesis has been reviewed extensively (4,18,51). ACh
biosynthesis is tightly linked to neuronal firing rate; choline and PC
administration are most effective in increasing ACh synthesis when
cholinergic neurons are discharging rapidly. This relationship was
initially described in the peripheral nervous system (15,118) and then
recognized in the central nervous system (71).

Choline circulates in the blood stream and crosses the blood-brain
barrier by a specific low affinity, unsaturated uptake system that is

sensitive to lithium (29). Exogenous choline enters nerve cells by a low affinity uptake system that is distinct from the high affinity uptake system that captures choline released into the synapse as a result of ACh hydrolysis. When plasma levels of choline are high, there is a net influx of choline into the brain; when plasma choline levels are low, there may an efflux of choline from brain to blood. Plasma choline levels normally vary between 5 and 15 nm/ml, depending upon the composition of the diet (126). Ingestion of choline salts increases plasma choline levels and may be expected to increase brain choline and brain ACh levels in humans as in rats; choline administration is known to increase CSF choline levels (45). PC is the preferred form of choline for precursor treatment: its administration produced higher and more prolonged plasma choline elevations than equimolar amounts of choline salts (124) and its ingestion did not impart the fishy body odor so characteristic of large doses of choline.

The amount of PC required to increase plasma choline levels, and thereby increase transport of choline into the brain, depends upon its purity. Using an 80% PC preparation (Unimills, Unilever), we determined that it was necessary to administer 5 to 7 grams in order to increase plasma choline levels significantly. Insufficient information exists regarding a therapeutic dose of PC. In most studies, enough PC has been given at least to double plasma choline levels, but in individual diseases this may be either too much (38,82) or too little (24). Human research with PC has been impeded by inadequate supplies of purified PC in a palatable form. Most commercially available lecithin preparations contain only 10 to 20% PC and much larger quantities of unknown phospholipids and possible impurities. At least 2 preparations are now available that contain purified PC in a palatable form: capsules containing 900 mg of PC (Advanced Nutritional Technology, Inc.) and PC-enriched packets of soup (T.J. Lipton Co.). We find that 9 capsules or 2 packets of instant soup are sufficient to double plasma choline levels (53).

RESULTS OF PC THERAPY IN BRAIN DISEASES

Based upon choline's ability to increase brain ACh levels in rats, Growdon et al. (44) postulated that pharmacological amounts of choline might be given to treat human diseases associated with deficient cholinergic tone. This hypothesis was called "dietary neurotransmitter precursor treatment" and has been tested in at least 8 neurological and psychiatric diseases. Precursor therapy has been most successful in tardive dyskinesia, and promising preliminary results have been obtained in treating affective disorders. Short-term administration of PC has not improved memory in Alzheimer's disease but its long-term administration, either as a single treatment or in combination with synthetic drugs, is currently under investigation. Equivocal or negative results have been reported with PC in the treatment of Parkinson's disease, Huntington's disease, Tourette Syndrome, amyotrophic lateral sclerosis, and the familial ataxias. In almost all clinical trials, PC has been well tolerated and free from toxic side effects (125).

Tardive Dyskinesia (TD). The efficacy of PC (and choline salts) in brain diseases is best documented in TD (42). TD is a hyperkinetic movement disorder characterized by buccal-lingual-masticatory choreiform movements that sometimes spread to involve limbs and trunk. TD is caused by antipsychotic drugs (39) and affects 10 to 15% of individuals receiving these medications (73). Although the pathophsyiology of TD remains uncertain (36), cholinergic mechanisms seem involved because anticholinergic drugs generally worsen and cholinergic agonists suppress the movements (6). Shortly after the initial report that choline increased ACh levels in rats, Davis et al. (30) gave choline chloride to a patient

122

with TD and reported a substantial decrease in choreiform movements; this result was subsequently confirmed in another 4 patients (31). In an independent study, Growdon et al. (52) gave choline chloride to 20 patients with persistent TD and found that the movements were suppressed in 9, temporarily worsened in 1, and unchanged in 10. Growdon et al. (50) then reported that PC partially suppressed TD movements in 4 additional patients. Similar positive results have been reported by numerous other investigators in both single and double blind studies (41,67,92,95). Complete suppression of TD is unusual and most investigators consider a 50% or greater reduction in dyskinetic movements a success. Not all TD patients benefit from PC: neither Branchey et al. (20) nor Domino et al. (34) found improvement with PC, although the counted number of dyskinetic movements decreased substantially in 4 of the 7 cases reported by Branchey et al. Possible reasons for variable responses include methodological problems such as interpatient variability of dyskinetic movements as well as the probable biochemical heterogeneity of individuals with the phenomenology of TD (43,106). Although no treatment is universally successful in suppressing persistent TD, PC is sufficiently safe and effective that it can be recommended as the first drug of choice (3,107).

Affective Disorders. The monoaminergic theory of affective disorders (72,85,108,109) has been modified to include cholinergic mechanisms (32,33). Support for cholinergic influences on mood come from reports that choline can induce depression in psychiatric patients with a depressive diathesis (113) and physostigmine can reduce acute manic symptoms (69,70). Based upon these pharmacological observations, Cohen et al. (24) gave 15 to 30 grams/day of 90% PC to 4 patients with acute manic symptoms. In addition to PC, 2 of the 4 patients received lithium and the other 2 received both lithium and a neuroleptic drug. All 4 patients improved; manic symptoms reappeared in 3 after PC was discontinued. In a subsequent trial of PC, Cohen et al. (23) found that manic symptoms improved significantly more with PC than with placebo in 5 of 6 patients treated. Although PC has not yet replaced lithium and neuroleptics in the treatment of acute mania, PC by itself apparently can reduce manic symptoms. Schreier described a 13 year old girl with mania who improved during treatment with 15 grams of 90% PC, whereas neuroleptics and lithium were ineffective (110).

Alzheimer's Disease (AD). This is a chronic progressive neurological disorder characterized by amnesia, aphasia, apraxia, personality changes, and behavioral abnormalities; primary motor and sensory modalities are relatively unaffected. The neuropathological changes in AD consist of neuronal loss in cortical regions (115) as well as in subcortical nuclei such as the nucleus basalis of Meynert (121), the locus coeruleus (19,86), and the raphe nuclei (66). Of the multiple neurotransmitter deficits that have been described, the reduction in ACh is most profound and the only one linked to dementia severity and extent of histopathological lesions (96,122). These pathological observations, coupled with pharmacological studies in normal subjects linking learning and memory to the cholinergic system (35), generated the cholinergic hypothesis of memory dysfunction in AD (27,47). Choline salts and PC have been given to patients with AD by more than 20 investigators and the results are uniformly negative: in short-term trials lasting 1 to 3 months, neither compound given alone produced improvement in memory or in any other cognitive function tested (27,28). There is a single report that PC given for 6 months may retard the progression of dementia in a subset of patients with AD (82) but confirmatory studies are necessary before accepting this result.

PC has been given in combination with synthetic drugs in order to potentiate their effects on cholinergic neurotransmission. Peters and

Levin (97) first noted that PC plus physostigmine improved memory in 5 patients with AD whereas neither drug alone was effective. Kaye et al. (75) reported similar results using PC in combination with another acetlcholinesterase inhibitor, THA. The most optimistic report was from Thal et al. (116) who found that 30 grams/day of 35% PC in combination with physostigmine increased memory by 23 to 43% and diminished intrusion errors by 43% in 6 of 7 patients. In contrast to these reports, Wettstein et al. (120) found no improvement with 18 grams/day of 98% PC in combination with physostigmine in 8 patients.

PC has also been administered in order to potentiate the effects of other drugs given to enhance memory in AD. Choline in combination with the nootropic piracetam was reported to improve mood in one study (40) and PC plus piracetam to improve some cognitive functions in another (111). In the largest series to date, Growdon et al. (49) detected no improvement in 18 patients with AD treated with 18 grams/day of 80% PC plus piracetam according to a double-blind protocol. In other studies, the joint administration of PC plus lithium (104) and PC plus hydergine (68) did not improve memory in patients with AD.

Parkinson's disease (PD). This is a degenerative neurological disorder with cardinal manifestations of akinesia, tremor, and rigidity. The basis for giving choline supplements to patients with PD rests upon 2 observations: the postulated balance between dopamine and acetylcholine neurotransmission in the striatum, and pathological evidence for atrophy of the cholinergic cells in the nucleus basalis of Meynert. In the first instance, the dopaminergic cells in the substantia nigra degenerate and there is diminished dopaminergic transmission in the striatum, with resultant diminished inhibition of cholinergic interneurons (63). Side effects of conventional treatment with dopaminergic drugs such as levodopa and bromocriptine often include dyskinetic movements. It was postulated that choline administration might suppress these involuntary movements without diminishing the beneficial effects of dopaminergic therapy (93). In 2 separate studies, choline chloride administration did in fact decrease levodopa-induced dyskinetic movements, but increased rigidity and akinesia (9,93). Although choline cannot be recommended as a practical treatment for levodopa-induced dyskinesias, these results provide the most direct clinical evidence that choline increases ACh neurotransmission in humans, and underscore the difficulty of restoring normal neurotransmitter balance within the striatum.

The prevalence of dementia in PD is highly controversial (48), and estimates of specific cognitive impairments approach 100 % of cases (102). Of the pathological lesions occurring in PD that may account for intellectual impairment, depletion of cholinergic neurons in the nucleus basalis of Meynert is a prime candidate (21). Barbeau (10) gave 20 grams/day of commercial PC to 10 demented Parkinsonians and noted a decrease in confusion and hallucinations; their performance on a test of constructional apraxia improved. In a similar study, Tweedy and Garcia(117) treated 16 elderly and mentally-impaired Parkinsonians with 32 grams/day of commercial PC. There were no overall benefits of PC treatment, although 1 patient improved slightly on a block design test. As a result of these studies, investigators have abandoned PC as a treatment for dementia in PD.

Huntington's disease (HD). This is a degenerative neurological disorder that is inherited as an autosomal dominant trait. The neurological manifestations of HD include personality change, dementia, involuntary movements (chorea), and poor balance and gait. The caudate nucleus is the site of major pathological change in HD; among the resultant neurochemical changes, a decrease in the marker for ACh is prominent (16,90). The pathological evidence of dminished ACh neurotransmission, and prior observations that physostigmine administration suppressed chorea (77), led investigators to test choline in HD. Davis et al. (31) treated 4 patients with choline chloride and reported benefit in 2. Aquilonius and Eckernas (5) reported a small decrease in choreic movements in 2 of 5 patients but concluded that overall choline did not significantly alter involuntary movements. In the largest series, Growdon et al. (46) gave choline chloride to 10 patients with HD and found no clinical improvement in any neurological symptom, despite significant elevations in plasma and CSF choline levels (45). Failure of choline administration to suppress chorea in the majority of patients with HD stands in contrast to its success in TD. It is unlikely that this difference is due to inadequate ACh synthesis but rather stems from the multiple neurotransmitter deficits now known to exist in this disorder (87). Even if choline or PC increased ACh neurotransmission in the caudate, the other biochemical abnormalities would remain.

Other Neurological Disorders. PC has been tested in amyotrophic lateral sclerosis (ALS), Tourette Syndrome (TS), and the familial spinocerebellar degenerative states; the results are either negative or inconclusive. ALS is a fatal disease in which there is specific degeneration of the cholinergic anterior horn cells whose axons project from the spinal cord to muscle. In order to increase ACh neurotranmission and restore muscle strength, Kelemen et al (76) gave 20 grams/day of 50% PC to 10 patients with ALS. There was no improvement in muscular function, despite treatment lasting 3 to 12 months.

TS is a neuropsychiatric disorder with motor and behavioral abnormalities. Symptoms include brief muscular spasms (convulsive tics), vocal tics such as grunts and barking sounds, and repetitive swearing. Most of these symptoms can be suppressed by neuroleptic drugs such as haloperidol (107). Evidence for a cholinergic deficit in TS stems from the observation that physostigmine can suppress tics in some patients (112,114). Barbeau (8) first reported some success with choline chloride in 1 of 3 cases of TS, and in a subsequent study (11) reported that 3 patients with TS improved with 40 to 50 grams/day of commercial lecithin. Polinsky and co-workers (103) could not confirm these results. They gave 35 grams/day of 55% PC to 6 patients with TS according to a double blind crossover protocol and found no improvement. Similar results were reported by Moldofsky and Sandor (91), who found no improvement in 5 patients treated with PC.

The group of degenerative diseases known as familial spinocerebellar degenerations share a common constellation of symptoms including ataxia, incoordination, spasticity, and gait impairments. The report that physostigmine administration produced some benefit in patients with Friedreich's ataxia (74) stimulated investigators to test ACh precursors in this disease. The suggestion that there was a defect in pyruvate dehydrogenase complex causing diminished ACh synthesis provided additional rationale for precursor therapy (12). The studies published to date report conflicting results, and the use of choline or PC in ataxic states remains inconclusive. Barbeau gave choline and PC (8,9) to patients with Friedreich's ataxia and patients with a spastic variant of spinocerebellar degeneration. He reported that 30 grams/day of commercial PC improved performance score in 10 patients with Friedreich's ataxia and seemed to slow the progression of symptoms. Neither choline nor PC improved the 6 patients with spastic ataxia. Several other case reports indicated improvement with choline chloride (81,83) and PC (37). In contrast to these reports, Lawrence et al. (80) conducted a placebo-controlled study of choline administration to 15 patients with a variety of ataxic diseases and found that choline administration significantly improved mobility and hand movement in only one. Three additional studies reported negative results in a total of 11 patients who received commercial PC according to double-blind crossover protocols (22,101,105). Pentland et al. (94) reported the largest series: Of 12 patients with Friedreich's ataxia treated with 25 grams/day of 96% PC according to a double-blind crossover protocol, none showed any consistent improvement during treatment. All of these studies taken together indicate that clinical benefits from PC administration were never impressive and generally non-existent; effective treatment for these disorders remains elusive.

PC AS AN INTEGRAL PART OF NEURONAL MEMBRANES

A second indication for PC administration in brain disorders recognizes the facts that PC is an integral part of the phospholipid (PL) component of neuronal membranes, and that alterations in the PL moiety may affect brain function by virtue of changing characteristics such as membrane fluidity (57) and neurotransmitter receptor number and responses (58). These observations provide a theoretical basis for administering PC but clinical studies testing this hypothesis have not begun. There are two overall postulates for PC administration: to correct PL abnormalities, and to stabilize neuronal membranes.

PC Given to Correct PL Abnormalities. Abnormalities in membrane composition and PL metabolism have been described in several tissues outside the central nervous system in humans, but few data exist regarding changes within the brain. In no instance has PC been given to correct any of the postulated abnormalities, although such treatment may be considered in the future.

Zubenko et al. (127) observed that membrane fluidity increased in platelets of patients with AD; this finding was a surprise because membrane fluidity generally decreases with age (56). In two independent studies, membrane fluidity was found to be decreased in acanthocytic red blood cells in patients with abetalipoproteinemia; there was an associated increase in sphingomyelin:PC ratios (13,64). Abnormalities in PC have been reported in major psychiatric disorders. Hitzeman et al.(61) found a significant decrease in membrane PC in erythrocytic membranes in patients with schizophrenic illnesses but not in patients with mania. In a subsequent study, they (62) ascribed the decrease in PC to decreased methylation enzyme activity. Similar results were reported by Alarcon et al. (1) in patients with depression. Changes in fatty acid composition of PC in erythrocytes from chronic alcoholics have been described and may contribute to low grade hemolysis. Alling et al. (2) found that polyunsaturated fatty acids decreased, and saturated and monosaturated fatty acids increased, in membranes from chronic alcoholics during withdrawal. Similar findings limited to PC were observed by LaDroitte et al. (79), who reported significant decreases in linoleic and increases in octadecenoic acids in patients with alcoholism. Abnormalities of PC and phosphatidylethanolamine metabolism have been reported in microsomal preparations of dystrophic muscle obtained from patients with Duchenne's dystrophy. Kunze et al. (78) found a decrease in PC 16:0 and 18:2 and increase in 18:0 and 18:1 fatty acids, as well as alterations in acylation of lysophosphatidylcholine and glycerol-3-phosphate. Based upon these and similar data, Infante (65) postulated that impaired synthesis of unsaturated PCs of the sarcoplasmic reticulum is the primary deficit in Duchenne's muscular dystrophy.

Attempts to collect direct evidence for PL abnormalities in brain disease are just beginning, and will be abetted by the development of nuclear magnetic resonance (NMR) spectroscopic techniques. NMR spectroscopy has been a major analytic chemical method for over 2 decades but its application to biological tissue is new, and should complement traditional biochemical methods for detecting PL abnormalities and effects of PC treatment. Among the various nuclei with potential for study, phosphorus 31 (^{31}P) has the greatest appeal because of overall interest in high energy phosphorus-containing compounds and in components of PL metabolism. Previous studies indicated that ^{31}P NMR in vitro spectroscopy could detect phosphomonoesters (PMs) such as phosphorylcholine and phosphorylethanolamine, and phosphodiesters (PDs) such as glycerophosphoroethanolamine (GPE) and glcerophospholcholine (GPC), and that their concentrations were relatively unaffected by ischemia, apoxia, or death. Three ^{31}P NMR in vitro spectroscopic studies have been conducted on aqueous solutions of brain tissue from patients with AD. Pettegrew et al. (98-100) analyzed ^{31}P NMR spectra based upon postmortem brain samples obtained from 3 patients: classic AD, atypical AD (with histopathological lesions limited to the hippocampus), and a normal control subject. The authors reported that in AD, phosphorylcholine and phosphorylethanolamine levels were twice as high as those in the normal subject, and levels of GPC and GPE were 1 1/2 times higher. These findings were interpreted as being consistent with an elevation of phospholipase activity in degenerating neurons. In a similar ^{31}P NMR spectroscopic study, Barany et al. (7) reported that the ratio of GPC:GPE was 0.28 in brains of 9 normal control subjects and 0.84 in 9 AD patients. In contrast to Pettegrew et al., Barany et al. found no change in phosphorylcholine nor phosphorylethanolamine levels in AD. Miatto et al. (88) performed ^{31}P NMR spectroscopy on brain samples from 7 patients with AD and 9 control subjects. They found evidence for altered PL metabolism in that relative levels of PMs were decreased and PDs increased in frontal and parietal regions in patients with AD compared to control subjects. These data are consistent with the hypothesis that abnormalities in PL metabolism contribute to possible neuronal membrane dysfunction and impaired cholinergic transmission in AD. Whether PC administration will correct these changes remains to be tested.

PC Given to Stabilize Neuronal Membranes. A second theoretical reason to administer PC is the hypothesis that its administration will preserve the structural and functional integrity of neurons, and prevent the neuronal dissolution that occurs in the course of normal aging or in neurodegenerative diseases such as AD (54,123). Preclinical experiments suggest that diets rich in choline may preserve neuronal architecture and retard toxin-induced or age-related decrements in animal behaviors.

Mizumori et al. (89) found that choline-enriched diets protected old mice from the amnesic effects of the drug anisomycin. Bartus et al. (14) studied retention of acquired passive avoidance behavior in mice ranging from 2 to 31 months old and found age-related decrements in test performance. In a second experiment, separate groups of mice received choline-deficient and choline-enhanced diets for 4 months and were retested in passive avoidance. There was the expected decline in performance with age among mice fed the choline-deficient diet, but not in those receiving choline enrichment. In fact, old mice maintained on the high choline diet performed as well as young 3 month old mice. In related anatomic studies, there was greater dendritic spine density on neocortical pyramidal neurons of mice receiving choline-enriched than choline-deficient diets. These studies suggest that the behavioral effects of long-term dietary supplementation with a choline source result from preserved synaptic structure. Clinical correlates of these animal studies are scarce, although the single report that long-term PC ingestion slowed cognitive decline in elderly patients with AD (82) supports this approach to treatment.

Another reason to believe that changes in brain membrane PL composition will be observed in AD and might contribute to its pathology stems from the observation that neuronal PC serves as a reservoir for choline that is available for ACh synthesis (18). The fact that cholinergic neurons are especially sensitive to degeneration in AD may derive from the dual uses of choline: for incorporation into PC for neuronal membranes, and for synthesis to ACh (17,123). If the neuron selectively hydrolyses PC to yield choline for ACh synthesis, leaving other membrane phosphatides intact, this process would be expected to alter membrane composition and function radically. If, instead, the breakdown of PL caused compensatory reduction in the synthesis of other PLs or accelerated their breakdown, this condition might impair membrane remodelling believed to underlie certain aspects of learning and memory. Either situation would be expected to contribute to the accelerated degeneration in some of the long-axon cholinergic neurons that appear to be especially vulnerable in AD. The goal of chronic dietary supplements with PC would be to provide enough choline to support heightened demand for ACh synthesis without altering the chemical composition of neuronal membranes.

REFERENCES

1. Alarcon, R.D., Tolbert, L.C., Monti, J.A., Morere, D.A., Walter-Ryan, W.G., Kemp, B., Smythies, J.R., 1985, One-carbon metabolism disturbances in affective disorders. A preliminary report, J. Affective Disord., 9:297-301.

2. Alling, C., Gustavsson, L., Kristensson-Aas, A., Wallerstedt, S., 1984, Changes in fatty acid composition of major glycerophospholipids in erythrocyte membranes from chronic alcoholics during withdrawal, Scand. J. Clin. Lab. Invest., 44:283-289.

3. Ansell, G.B., 1981, The biochemical background to tardive dyskinesia, Neuropharmacology, 4:311-317.

4. Ansell, G.B., Spanner, S., 1978, The source of choline for acetylcholine synthesis, in: "Cholinergic Mechanisms and Psychopharmacology", D.J. Jenden, ed., Plenum, New York, pp.431-445.

5. Aquilonius, S.M., Eckernas, S.A., 1977, Choline therapy in Huntington's chorea, Neurology, 27:887-889.

6. Baldessarini, R.J., Tarsy, D., 1980, Dopamine and the pathophysiology of dyskinesias induced by antipsychotic drugs, Ann. Rev. Neurosci., 3:23-41.

7. Barany, M., Chang, Y.C., Arus, C., Rustan, T., Frey, W.H.II, 1985, Increased glycerol-3-phosphorylcholine in post-mortem Alzheimer's brain, Lancet i, 517.

8. Barbeau, A., 1978, Emerging treatments: replacement therapy with choline or lecithin in neurological diseases, Can. J. Neurol. Sci., 1:157-160.

9. Barbeau, A., 1979, Lecithin in movement disorders, in:"Choline and Lecithin in Brain Disorders", vol. 5, Nutrition and the Brain, A. Barbeau, J.H. Growdon, R.J. Wurtman, eds., Raven Press, New York, pp. 263-271.

10. Barbeau, A., 1980, Lecithin in Parkinson's disease, J. Neurol. Transm., 16:187-193.

11. Barbeau, A., 1980, Cholinergic treatment in the Tourette syndrome, N. Engl. J. Med., 302:1310-1311.

12. Barbeau, A., Butterworth, R.F., Ngo, T., et al., 1976, Pyruvate metabolism in Friedreich's ataxia, Can. J. Neurol. Sci., 3:379-388.

13. Barenholz, Y., Yechiel, E., Cohen, R., Deckelbaum, R.J., 1981, Importance of cholesterol-phospholipid interaction in determining dynamics of normal and abetalipoproteinemia red blood cell membrane, Cell Biophys., 3:115-126.

14. Bartus, R.T., Dean, R.L., Goas, J.A., Lippa, A.S., 1980, Age-related changes in passive avoidance retention: modulation with dietary choline, Science, 209:301-303.

15. Bierkamper, G.G., Goldberg, A.M., 1980, Release of acetylcholine from the vascular perfused rat phrenic nerve-hemidiaphragm, Brain Res., 202:234-237.

16. Bird, E.D., Iversen, L.L., 1974, Huntington's chorea: post-mortem measurement of glutamic acid decarboxylase, choline acetyltransferase and dopamine in basal ganglia, Brain, 97:457-472.

17. Blusztajn, J.K., Wurtman, R.J., 1981, Choline biosynthesis by a preparation enriched in synaptosomes from rat brain, Nature, 290:417-418.

18. Blusztajn, J.K., Wurtman, R.J., 1983, Choline and cholinergic neurons, Science, 221:614-620.

19. Bondareff, W., Mountjoy, C.Q., Roth, M., 1982, Loss of neurons of origin of the adrenergic projection of cerebral cortex (nucleus locus coeruleus) in senile dementia, Neurology, 32:164-168.

20. Branchey, M.H., Branchey, L.B., Bark, N.M., Richardson, M.A., 1979, Lecithin in the treatment of tardive dyskinesia, Commun. Psychopharmacol., 3:303-307.

21. Candy, J.M., Perry, R.H., Perry, E.H., Irving, D., Blessed, G., Fairbairn, A.D., Tomlinson, B.E., 1983, Pathological changes in the nucleus of Meynert in Alzheimer's and Parkinson's diseases, J. Neurol. Sci., 59:277-289.

22. Chamberlain, S., Robinson, N., Walker, J., Smith, C., Benton, S., Kennard, C., Swash, M., Kilkenny, B., Bradbury, S., 1980, Effect of lecithin on disability and plasma free-choline levels in Friedreich's ataxia, J. Neurol. Neurosurg. Psychiatry, 43:843-845.

23. Cohen, B.M., Lipinski, J.F., Altesman, R.I., 1982, Lecithin in the treatment of mania: double-blind, placebo-controlled trials, Am. J. Psychiatry, 139:1162-1164.

24. Cohen, B.M., Miller, A.L., Lipinski, J.F., Pope, H.G., 1980, Lecithin in mania: a preliminary report. Am. J. Psychiatry, 137:242-243.

25. Cohen, E.L., Wurtman, R.J., 1975, Brain acetylcholine: increase after systemic choline administration, Life Sci., 16:1095-1102.

26. Cohen E.L., Wurtman, R.J., 1976, Brain acetylcholine: control by dietary choline, Science, 191:561-562.

27. Corkin, S., 1981, Brain acetylcholine, aging, and Alzheimer's disease: implications for treatment, Trends in Neuroscience, 4:287-290.

28. Corkin, S., Davis, K., Growdon, J., Usdin, E., Wurtman, R., 1982, Alzheimer's Disease: A Report of Progress in Research, Raven Press, New York.

29. Cornford, E.M., Braun, L.D., Oldendorf, W.H., 1978, Carrier mediated blood-brain barrier transport of choline and certain choline analogs, J. Neurochem., 30: 299-308.

30. Davis, K.L., Berger, P.A., Hollister, L.E., 1975, Choline for tardive dyskinesia, N. Engl. J. Med., 293:152.

31. Davis, K.L., Hollister, L.E., Barchas, J.D., et al., 1976, Choline in tardive dyskinesia and Huntington's disease, Life Sci., 19: 1507-1519.

32. Davis, K.L., Hollister, L.E., Berger, P.A., Barchas, J.D., 1975, Cholinergic imbalance hypotheses of psychoses and movement disorders: strategies for evaluation, Psychopharmacol. Commun., 1:533-543.

33. Davis, J.M., 1975, Critique of single amine theories: evidence of a cholinergic influence in the major mental illnesses, Res. Publ. Assoc. Res. Ment. Dis., 54:333-346.

34. Domino, E.F., May, W.W., Demetriou, S., Mathews, B., Tait, S., Kovacic, B., 1985, Lack of clinically significant improvement of patients with tardive dyskinesia following phosphatidylcholine therapy, Biol. Psychiatry, 20:1189-1196.

35. Drachman, D.A., 1977, Memory and cognitive function in man: Does the cholinergic system have a specific role? Neurology, 27:783-790.

36. Fibiger, H.C., Lloyd, K.G., 1984, Neurobiological substrates of tardive dyskinesia: the GABA hypothesis, Trends in Neurosci., 7:462-464.

37. Filla, A., Campanella, G., 1982, A six-month phosphatidylcholine trial in Friedreich's ataxia, Can. J. Neurol. Sci., 9:147-150.

38. Finocchiaro, G., DiDonato, S., Madonna, M., Fusi, R., Ladinsky, H., Consolo, S., 1985, An approach using lecithin treatment for olivopontocerebellar atrophies, Eur. Neurol., 24:414-421.

39. Food and Drug Administration Task Force, American College of Neuropsycho- pharmacology, 1973, Neurological syndromes associated with antipsychotic drug use: a special report, Arch. Gen. Psychiatry, 28:463-467.

40. Friedman E., Sherman, K., Ferris, S.H., Reisberg, B., Bartus, R.T., Schenk, M.K., 1981, Clinical response to choline plus piracetam in senile dementia: relation to red cell choline levels, N. Engl. J. Med., 304:1490-1491.

41. Gelenberg, A.J., Doller-Wojcik, J.C., Growdon, J.H., 1979, Choline and lecithin in the treatment of tardive dyskinesia: preliminary results from a pilot study, Am. J. Psychiatry, 136:772-776.

42. Gibson, G.E., Blass, J.P., 1976, Impaired synthesis of acetylcholine in brain accompanying mild hypoxia and hypoglycemia, J. Neurochem., 27:37-42.

43. Growdon, J.H., 1986, Phosphatidylcholine and tardive dyskinesia (letter), Biol. Psychiatry, 21:702-703.

44. Growdon, J.H., Cohen, E.L., Wurtman, R.J., 1977, Treatment of brain diseases with dietary precursors of neurotransmitters, Ann. Intern. Med., 86:337-339.

45. Growdon, J.H., Cohen, E.L., Wurtman, R.J., 1977, Effects of oral choline administration on serum and CSF choline levels in patients with Huntington's disease, J. Neurochem., 28:229-231.

46. Growdon, J.H., Cohen, E.L., Wurtman, R.J., 1977, Huntington's disease: clinical and chemical effects of choline administration, Ann. Neurol., 1:418-422.

47. Growdon, J.H., Corkin, S., 1979, Neurochemical approaches to the treatment of senile dementia, in:"Psychopathology in the Aged", J.O. Cole, J. Barrett, eds., Raven Press, New York, pp. 281-294.

48. Growdon, J.H., Corkin, S., in press, Cognitive impairments in Parkinson's disease, in:"Parkinson's Disease (Advances in Neurology series)", M.D. Yahr, R.C. Duvoisin, eds., Raven Press, New York.

49. Growdon, J.H., Corkin, S., Huff, F.J., Rosen, T.J., 1986, Piracetam combined with lecithin in the treatment of Alzheimer's disease, Neurobiol. Aging, 7:269-276.

50. Growdon, J.H., Gelenberg, A.J., Doller, J., Hirsch, M.J., Wurtman, R.J., 1978, Lecithin can suppress tardive dyskinesia. N. Engl. J. Med., 298:1029-1030.

51. Growdon, J.H., Gibson, C.J., 1982, Dietary precursors of neurotransmitters: treatment strategies, in:"Current Neurology", S.H. Appel, ed., Wiley and Sons, New York, pp. 117-144.

52. Growdon, J.H., Hirsch, M.J., Wurtman, R.J., Weiner, W., 1977, Oral choline administration to patients with tardive dyskinesia, N. Engl. J. Med., 297:524-527.

53. Growdon, J.H., Wheeler, S., Graham, H.N., 1984, Plasma choline responses to lecithin-enriched soup, Psychopharmacol. Bull., 20:603-606.

54. Growdon, J.H., Wurtman, R.J., 1983, The future of cholinergic precursor treatment in Alzheimer's disease, in:"Banbury Report 15: Biological Aspect of Alzheimer's Disease", L. Moran, ed., Banbury Center, Cold Spring Laboratory, pp. 451-459.

55. Haubrich, D.R., Wang, P.F.L., Clody, D.E., Wedeking, P.W., 1975, Increases in rat brain acetylcholine induced by choline or deanol, Life Sci., 17:975-980.

56. Hegner, D., 1980, Age dependence of molecular and functional changes in biological membrane properties, Mech. Aging Develop., 14:101-118.

57. Hirata, F., Axelrod, J., 1978, Enzymatic methylation of

phosphatidylethanolamine increases erythrocyte membrane fluidity, Nature, 275:219-220.

58.   Hirata, F., Axelrod, J., 1980, Phospholipid methylation and biological signal transmission, Science, 209:1082-1090.

59.   Hirsch, M.J., Growdon, J.H., Wurtman, R.J., 1977, Increase in hippocampal acetylcholine after choline administration, Brain Res., 125:383-385.

60.   Hirsch, M.J., Wurtman, R.J., 1978, Lecithin comsumption elevates acetylcholine concentrations in rat brain and adrenal gland, Science, 207:223-225.

61.   Hitzemann, R., Hirschowitz, J., Garver, D., 1984, Membrane abnormalities in the psychoses and affective disorders, J. Psychiatr. Res., 18:319-326.

62.   Hitzemann, R., Mark, C., Hirschowitz, J., Garver, D., 1985, Characteristics of pohspholipid methylation in human erythrocyte ghosts: relationship(s) to the psychoses and affective disorders, Biol. Psychiatry, 20:397-407.

63.   Hornykiewicz, O., 1982, Brain neurotransmitter changes in Parkinson's disease, in:"Movement Disorders", C.D. Marsden, S. Fahn, eds., Butterworth Scientific, London, pp. 41-58.

64.   Iida, H., Takashima, Y., Maeda, S., Sekiya, T., Kawade, M., Kawamura, M., Okano, Y., Nozawa, Y., 1984, Alterations in erythrocyte membrane lipids in abetalipoproteinemia: phospholipid and fatty acyl composition, Biochem. Med., 32:79-87.

65.   Infante, J.P., 1986, Defective synthesis of polyunsaturated phosphatidylcholines as the primary lesion in Duchenne and murine dy muscular dystrophies, Med. Hypotheses, 19:113-116.

66.   Ishi, T., 1966, Distribution of Alzheimer's neurofibrillary changes in the brainstem and hypothalamus of senile dementia, Acta Neurol. Pathol, 6:181-187.

67.   Jackson, I.V., Nuttall, E.A., Ibe, I.O., Perez-Cruet, J., 1979, Treatment of tardive dyskinesia with lecithin, Am. J. Psychiatry, 136:1458-1460.

68.   Jenike, M.A., Albert, M.S., Heller, H., LoCastro, S., Gunther, J., 1986, Combination therapy with lecithin and ergoloid mesylates for Alzheimer's disease, J. Clin. Psychiatry, 47:249-251.

69.   Janowsky, D.S., El-Yousef, M.K., Davis, J.M., Hubbard, H., Sekerke, H.J., 1972, Cholinergic reversal of manic symptoms, Lancet, 1:1236-1237.

70.   Janowsky, D.S., El-Yousef, K., Davis, J.M., Sekerke, H.J., 1973, Parasympathetic suppression of manic symptoms by physostigmine, Arch. Gen. Psychiatry, 28:542-547.

71.   Jenden, D.J., Weiler, M.H., Gundersen, C.B., 1982, Choline availability and acetylcholine synthesis, in: " Alzheimer's Disease: A Report of Progress in Research", S. Corkin, K.L. Davis, J.H. Growdon, et al., eds., Raven Press, New York, pp. 315-326.

72. Jones, F.D., Mass, J.W., Dekirmenjian, H., et al., 1975, Diagnostic subgroups of affective disorders and their urinary excretion of catecholamine metabolites, Am. J. Psychiatry, 132:1141-1148.

73. Kane, J.M., Woerner, M., Borenstein, M., Wegner, J., Lieberman, J., 1986, Integrating incidence and prevalence of tardive dyskinesia, Psychopharmacol. Bull., 22:254-258.

74. Kark, P.A., Blass, J., Spence, M.A., 1977, Physostigmine in familial ataxias, Neurology, 27:70-72.

75. Kaye, W.H., Sitaram, N., Weingartner, H., Ebert, M.H., Smallberg, S., Gillin, J.C., 1982, Modest facilitation on memory in dementia with combined lecithin and anticholinesterase treatment, Biol. Psychiatry, 17:275-280.

76. Kelemen, J., Hedlund, W., Murray-Douglas, P., Munsat, T.L., 1982, Lecithin is not effective in amyotrophic lateral sclerosis, Neurology, 32:315-316.

77. Klawans, H.L., Jr., Rubovits, R., 1972, Central cholinergic-anticholinergic antagonism in Huntington's chorea, Neurology, 22:107-112

78. Kunze, D., Rustow, B., Kuksis, A., Myher, J.J., 1986, Acylation of lysophosphatidylcholine and glycerolphosphate and fatty acid pattern in phosphatidylcholine and ethanolamine in microsomes of normal and dystrophic human muscle, Acta Neurol. Scand., 73:125-130.

79. LaDroitte, P., Lamboeuf, Y., de Saint Blanquat, G., Bezaury, J.P., 1985, Sensitivity of individual erythrocyte membrane phospholipids to changes in fatty acid composition in chronic alcoholic patients, Alcoholism (NY), 9:135-137.

80. Lawrence, G.M., Millac, P., Stout, G.S., et al., 1980, The use of choline chloride in ataxic disorders, J. Neurol. Neurosurg., Psychiatry, 43:452-545.

81. Legg, N.J., 1978, Oral choline in cerebellar ataxia, Br. Med. J., 2:1403-1404.

82. Little, A., Levy, R., Chuaqui-Kidd, P., Hand, D., 1985, A double-blind, placebo controlled trial of high-dose lecithin in Alzheimer's disease, J. Neurol. Neurosurg., Psychiatry, 48:736-742.

83. Livingstone, I.R., Mastaglia, F.L., 1979, Choline chloride in the treatment of ataxia, Br. Med. J., 2:939.

84. London, E.D., Coyle, J.T., 1978, Pharmacological augmentation of acetylcholine levels in kainate lesioned rat striatum, Biochem. Pharmacol., 27:2962-2965.

85. Maas, J.W., 1975, Biogenic amines and depression - biochemical and pharmacological separation of two types of depression, Arch. Gen. Psychiatry, 32:1357-1361.

86. Mann, D.M.A., Lincoln, J., Yates, P.D., et al., 1980, Changes in monoamine containing neurons of the human CNS in senile dementia, Brit. J. Psychiatry, 136:533-541.

87.   Martin, J.B., 1984, Huntington's disease: new approaches to an old problem, Neurology, 34:1059-1072.

88.   Miatto, O., Gonzalez, R.G., Buonanno, F., Growdon, J.H., in press, In vitro ^{31}P NMR spectroscopy detects altered phospholipid metabolism in Alzheimer's disease, Can. J. Neurol. Sci.

89.   Mizumori, S.J.Y., Patterson, T.A., Sternberg, H., Rosenzweig, M.R., Bennett, E.L., Timiras, P.S., 1985, Effects of dietary choline on memory and brain chmistry in aged mice, Neurobiol. of Aging, 6:51-56.

90.   McGeer, P.L., McGeer, E.G., Fibiger, H.C., 1973, Choline acetylase and glutamic acid decarboxylase in Huntington's chorea, Neurology, 23:912-917.

91.   Moldofsky, H., Sandor, P., 1983, Lecithin in the treatmen of Gilles de la Tourette's syndrome, Am. J. Psychiatry, 140:1627-1629.

92.   Newhouse, P., Bridenbaugh, R.H., 1981, Pharmacologic characterization and lecithin treatment of a patient with spontaneous oral-facial dyskinesia and dementia. Am. J. Psychiatry, 138:251-252.

93.   Papavasiliou, P.S., Rosal, V., 1979, Effects of choline in patients with levodopa-induced dyskinesia, in: "Choline and Lecithin in Brain Disorders", A. Barbeau, J.H. Growdon, R.J. Wurtman, eds., Raven Press, New York, pp. 335-341.

94.   Pentland, B., Martyn, C.N., Steer, C.R., Christie, J.E., 1981, Lecithin treatment in Friedreich's ataxia, Br. Med. J. (Clin. Res.), 282:1197-1198.

95.   Perez-Cruet, J., Menendez, I., Alvarez-Ghersi, J., Falcon, J.R., Valderrabano, O., Castro-Urrutia, E.C., Ifarraguerri, C., Perez, L.L., 1981, Double-blind study of lecithin in the treatment of persistent tardive dyskinesia. Bol. Assoc. Med. PR, 73:531-537.

96.   Perry, E.K., Tomlinson, B.E., Blessed, G., Bergmann, K., Gibson, P.M., Perry, R.H., 1978, Correlation of cholinergic abnormalities with senile plaques and mental scores in senile dementia, Brit. Med. J., 2:1457-1459.

97.   Peters, B.H., Levin, H.S., 1979, Effects of physostigmine and lecithin on memory in Alzheimer's disease, Ann. Neurol., 6:219-221.

98.   Pettegrew, J.W., Minshew, N.J., Cohen, M.M., Kopp, S.J., Glonek, T., 1984, 31-P changes in Alzheimer's and Huntington's disease brain, Neurology, 34:281.

99.   Pettegrew J.W., Minshew, N., Glonek, T., Kopp, S., Cohen, M., 1982, 31-P nuclear magnetic resonance analysis of Huntington and control brain, Neurology, 32:196.

100.   Pettegrew, J.W., Minshew, N.J., Diehl, J., Smith, T., Kopp, S.J., Glonek, T., 1983, Anatomical considerations for interpreting topical P-31 NMR, Lancet, ii, 913.

101.   Philcox, D.V., Kies, B., 1979, Choline in hereditary ataxia, Brit. Med. J., 2:613.

102.   Pirozzolo, F.J., Hansch, E.C., Mortimer, J.A., Webster, D.D., Kuskowski,

M.A., 1982, Dementia in Parkinson's disease: a neuropsychological analysis, Brain and Cognition, 1:71-83.

103. Polinsky, R.J., Ebert, M.H., Caine, E.D., Ludlow, C., Bassich, C.J., 1980, Cholinergic treatment in the Tourette syndrome, N. Engl. J. Med., 302:1310.

104. Randels, P.M., Marco, L.A., Ford, D.I., Mitchell, R., Scholl, M., Plesnarski, J., 1984, Lithium and lecithin treatment in Alzheimer's disease: a pilot study, Hillside J. Clin. Psychiatry, 6:139-147.

105. Reding, M.J., Blass, J.P., Stern, P.H., et al., 1981, Lecithin in hereditary ataxia, Neurology, 31:363-364.

106. Richardson, M.A., Craig, T.J., Branchey, M.H., 1982, Intra-patient variability in the measurement of tardive dyskinesia, Psychopharmacology (Berlin), 76:269-272.

107. Scheife, R.T., Growdon, J.H., 1982, Treating tardive dyskinesia, Seminars in Neurology, 2:305-315.

108. Schildkraut, J.J., 1965, The catecholamine hypothesis of affective disorders: a review of supporting evidence, Am. J. Psychiatry, 122:509-522.

109. Schildkraut, J.J., Orsulak, P.J., Schatzberg, A.F., et al., 1978, Toward a biochemical classification of depressive disorders, Part I, Arch. Gen. Psychiatry, 35:1427-1433.

110. Schreier, H.A., 1982, Mania responsive to lecithin in a 13-year-old girl, Am. J. Psychiatry, 139:108-110.

111. Smith, R.C., Vroulis, G., Johnson, R., Morgan, R., 1984, Comparison of therapeutic response to long-term treatment with lecithin versus piracetam plus lecithin in patients with Alzheimer's disease, Psychopharmacol. Bull., 20:542-545.

112. Stahl, S.M., Berger, P.A., 1980, Cholinergic treatment in the Tourette syndrome, N. Engl. J. Med., 302:1311.

113. Tamminga, C.A., Smith, R.C., Erickson, S.E., et al., 1977, Cholinergic influences in tardive dyskinesia, Am. J. Psychiatry, 134:769-774.

114. Tanner, C.M., Goetz, C.G., Klawans, H.L., 1982, Cholinergic mechanisms in Tourette syndrome, Neurology, 32:1315-1317.

115. Terry, R.D., Peck, A., DeTeresa, R., Schechter, R, Horoupian, D.S., 1981, Some morphometric aspects of the brain in senile dementia of the Alzheimer type, Ann. Neurol., 10:184-192.

116. Thal, L.J., Fuld, P.A., Masur, D.M., Sharpless, N.S., 1983, Oral physostigmine and lecithin improve memory in Alzheimer's disease, Ann. Neurol., 13:491-496.

117. Tweedy, J.R., Garcia, C.A., 1982, Lecithin treatment of cognitively impaired Parkinson's patients, Eur. J. Clin. Invest., 12:87-90.

118. Ulus, I.H., Hirsch, M.J., Wurtman, R.J., 1977, Trans-synaptic induction of adrenal tyrosine hydroxylase activity by choline: evidence that choline administration increases cholinergic transmission, Proc. Natl. Acad. Sci. USA., 74:798-800.

119. Wecker, L., Dettbarn, W.D., Schmidt, E.D., 1978, Choline administration: modification of the central actions of atropine, Science, 199:86–87.

120. Wettstein, A., 1983, No effect from double-blind trial of physostigmine and lecithin in Alzheimer disease, Ann. Neurol., 13:210–212.

121. Whitehouse, P.J., Price, D.L., Coyle, J.T., DeLong, M.R., 1981, Alzheimer's disease: evidence for selective loss of cholinergic neurons in the nucleus basalis, Ann. Neurol., 10:122–126.

122. Wilcock, G.K., Esiri, M.M., Bowen, D.M., Smith, C.C.T., 1982, Alzheimer's disease: correlation of cortical choline acetyltransferase activity with the severity of dementia and histological abnormalities, J. Neurol. Sci., 57:407–417.

123. Wurtman, R.J., Blusztajn, J.K., Maire, J.C., 1985, "Autocannibalism" of choline-containing membrane phospholipids in the pathogenesis of Alzheimer's disease – a hypothesis, Neurochem. Int., 7:369–372.

124. Wurtman, R.J., Hirsch, M.J., Growdon, J.H., 1977, Lecithin consumption raises serum free choline levels, Lancet, 2:68–69.

125. Yackulic, C.F., Anderson, B.G., Reker, D., Webb, E., Volavka, J., 1982, The safety of lecithin diet supplementation in schizophrenic patients, Biol. Psychiatry, 17:1445–1448.

126. Zeisel, S.H., Growdon, J.H., Wurtman, R.J., Magil, S.G., Logue, M., 1980, Normal plasma choline responses to ingested lecithin, Neurology, 30:1226–1229.

127. Zubenko, G.S., Cohen, B.M., Growdon, J.H., Corkin, S., 1984, Cell membrane abnormality in patients with Alzheimer's disease, Lancet, 2:235.

THE THERAPEUTIC VALUE OF PHOSPHATIDYLSERINE EFFECT IN THE AGING BRAIN

Gino Toffano

Fidia Neurobiological Research Laboratories
Via Ponte della Fabbrica, 3/A
35031 Abano Terme, Italy

INTRODUCTION

The Principles of Phospholipid Pharmacology

As the elucidation of the biological roles of membrane phospholipids has progressed, a new field of investigation dealing with the pharmacological effects of these compounds has become possible (Table 1). Cellular membranes in addition to their property of delimiting compartments, are endowed with surface recognition sites which, in specialized instances, yield the means of offense and defense reactions (Bruni and Toffano, 1982). The administration of phospholipid liposomes has been shown to activate blood clotting (Zwaal, 1978), complement pathways (Cunningham et al., 1979; Richards et al., 1979) and phagocytosis (Hudson et al., 1979; Wu et al., 1981). In addition, the exogenous phospholipids may enter the pathways of phospholipid metabolism yielding a number of short-lived compounds showing powerful regulation effects on biological and physiopathological phenomena. These products can be named lipid autacoid drugs since they are active drugs made in the organism itself.

Table 1

PHARMACOLOGICAL PROPERTIES OF THE PHOSPHOLIPID BILAYER

BIOLOGICAL FUNCTION	DERIVED PHARMACOLOGICAL EFFECT
- Barrier to hydrophilic compounds	- Transport of polar drugs
- Incorporation of hydrophobic or amphipathic compounds	- Transport of apolar or amphipathic drugs
- Display of functional groups at the bilayer surface	- Surface dependent effects
- Bilayer-bilayer fusion	- Release of entrapped material inside the cells
- Source of lipid chemical mediators	- Autacoid effects

(for a review see Bruni and Palatini, 1982)

Knowledge of the mechanism of action of the phospholipid bilayer enables to a certain degree to regulate its final effect. Hence, administration of exogenous phospholipids may lead to restoration of those cell functions which depend on membrane properties and on proper membrane turnover and cycling. The restoration phenomenon may be particularly evident at the cerebral level in view of the dependence of brain function on the integrity of various membrane structures.

Physiological and Pharmacological properties of phosphatidylserine

Phosphatidylserine (PtdSer) is the major acid phospholipid of eukaryotic cells. Its peculiar influence on the membrane stability and organization is linked to the properties of the serine headgroup, for a review see Bruni and Palatini (1982). Since the first observation of Verkleij et al. (1973) and Gordesky and Marinetti (1973) it is becoming increasingly manifest that negatively charged phospholipids, chiefly PtdSer, are predominantly distributed in the inner side of the plasma membrane in eukaryotic cells. The mechanism underlying this asymmetric distribution is not clear but a recent hypothesis based on the interaction of the negatively charged headgroup with the elements of cytoskeleton seems very attractive (Franck et al., 1983). If this mechanism is correct it can explain also the experimental observations (reviewed in Op den Kamp, 1979) according to which PtdSer is distributed outside the membrane surrounding the intracellular organelles. The opposite distribution asymmetry of PtdSer in the plasma membrane and in the membrane of intracellular organelles allows the polar head of this phospholipid to come into contact, when the intracellular traffic leads those internal structures to reach the inner face of the plasma membrane. Detailed analyses on the fusogenic properties of PtdSer have been performed some years ago (Papahadjopoulos, 1978; Sundler et al., 1981). The fusogenic property of PtdSer is believed to play a role in the process of exocytosis as this phospholipid is found in the inner side of plasma membrane and in the outer side of the intracellular vesicles transporting the material to be exported. Considering the pharmacological aspect, the fusogenic property of PdtSer becomes important when vesicles of this phospholipid are injected and may be incorporated in the plasma membrane of other cells by fusion. Among the lipid-dependent enzymes which are optimally activated by PtdSer, two of them deserve special mention. The first is the sodium-potassium-stimulated ATPase (Palatini et al., 1972). This enzyme owes its importance to the role of maintaining the gradient between the intracellular and the extracellular concentration of sodium and potassium. A further PtdSer-dependent enzyme deserving attention is the recently discovered (Castagna et al., 1982) protein kinase C. Its role on cell activation (Nishizuka, 1984) and its distribution in the central nervous system (Murphy et al., 1983) have been recently reported.

Exogenous phosphatidylserine, extracted and purified from bovine brain (BC-PS), has been shown to modify both neurotransmission and related behaviour in young-adult animals (Toffano and Bruni, 1980; Zanotti et al., 1984). BC-PS administration stimulates tyrosine-hydroxylase (TH) activity in rat brain (Toffano et al., 1978), induces dopamine (DA) release from rat striatal dopaminergic terminals (Mazzari and Battistella, 1980) and activates adenylate-cyclase in rat hypothalamus (Toffano et al., 1978; Calderini et al., 1981). Furthermore BC-PS stimulates acetylcholine (Ach) output from the cerebral cortex (Casamenti et al., 1979), and partially antagonizes the scopolamine-induced disruption of spontaneous alternation behaviour in a Y maze (Pepeu et al., 1980) and the decreased performances in passive avoidance behaviour induced by anoxia, scopolamine and electroshock (Zanotti et al., 1986). These effects prompted us to investigate the pharmacological potentialities of this phospholipid in the aging rat brain.

BC-PS TREATMENT RESTORES AGE-DEPENDENT ELECTROPHYSIOLOGICAL AND BEHAV-
IORAL ALTERATIONS IN RATS

Aging significantly reduces the efficiency of memory as indicated by
the decreased performance in the passive avoidance test (Fig. 1). The
phenomenon is not homogeneous since in male Sprague-Dawley rats the
decline of memory function appears only with the occurrence of abnormal,
spontaneous, asymptomatic spike-wave discharges in the EEG recordings
(Aporti et al., 1985). The abnormal EEG activity occurs in about 15% of
10-12 month-old rats and in 90% of 20-24 month-old rats. The EEG pattern
permits to identify a subpopulation of middle-aged rats which develop an
early decline of memory, where the extent of memory deficit seems to
correlate with the number of bursts present in the EEG recordings.

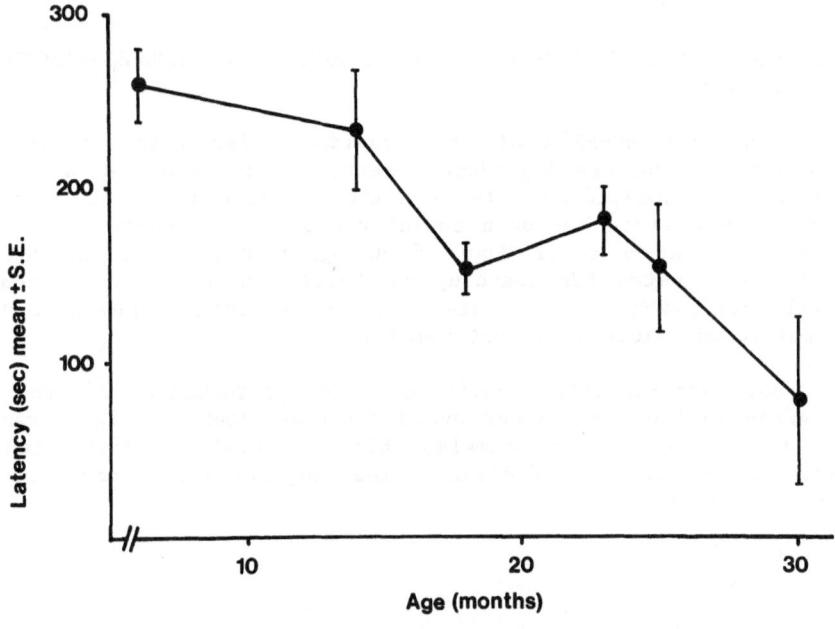

Fig. 1.

Retention of a passive avoidance response in rats of different age.

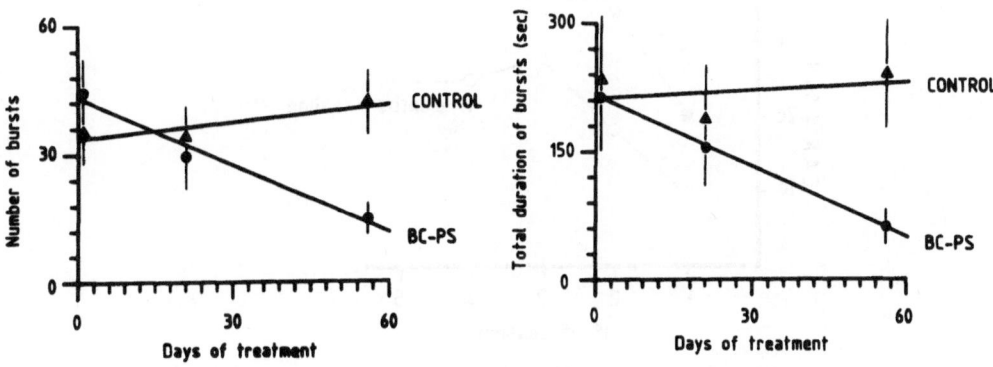

Fig. 2.

Effect of BC-PS treatment on the number and total duration of bursts in
rats of 17 months. Rats were daily injected with BC-PS at 15 mg/kg/ml
i.p. Control group received 50 mm Tris-HCl pH 7.5, 1 ml/kg i.p. For the
methodology see Aporti et al., 1985.

In this group of rats, prolonged treatment with 15 mg/kg/ip BC-PS significantly reduces both the number and the duration of abnormal spikes (Fig. 2). Other phospholipids such as phosphatidylcholine (PC), or the phosphatidylserine constituents, namely serine and oleic acid, do not possess a similar pharmacological activity, thus suggesting a structure-activity relationship (Calderini et al., 1986). BC-PS treatment leads also to a concomitant improvement of behavioural performance in the passive avoidance test suggesting a recovery of memory function. Increased retention of passive avoidance test after phosphatidylserine treatment has been previously observed in aged Wistar (Drago et al., 1981) and Fischer 344 (Corwin et al., 1985) rat strain. The capability of restoring cognitive deficit has been correlated with the capability of BC-PS to restore the decreased release of ACh in electrically stimulated cortical slices (Pedata et al., 1985).

BC-PS TREATMENT PREVENTS AGE-DEPENDENT OCCURRENCE OF MORPHO-FUNCTIONAL ALTERATIONS IN RATS

To know whether BC-PS could be provided also with preventive effects on some of the age-dependent cerebral dysfunctions we have administered the phospholipid to rats from the age of 4 months until about 2 years. The phospholipid has been administered by oral route, dissolved in the tap water, at a total dose of 50 mg/kg/day. At the age of 19 months, rats were tested for learning capability in an active avoidance test (shuttle box), while at the age of 22 – 24 months, they were used for morphometric and biochemical determinations.

Under our experimental conditions the performance of the 19 month-old group in the conditioned avoidance test (CAR) is significantly lower than that of the young animals, while the oral administration of BC-PS completely prevents the decline of learning capability occurring in the aged rats (Fig. 3).

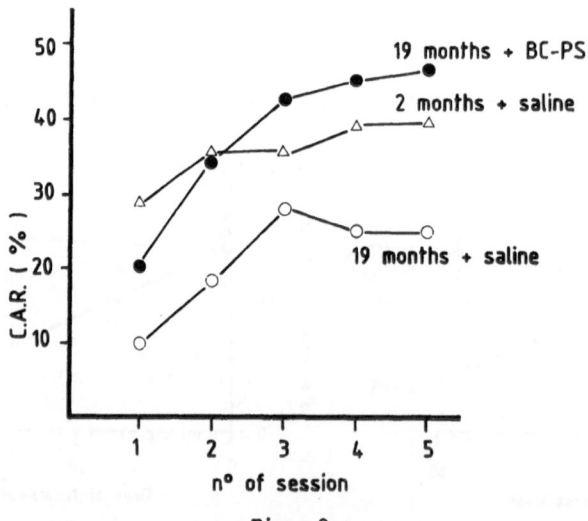

Fig. 3.

Effect of BC-PS treatment on active avoidance response of old rats. BC-PS was given orally at a daily dose of 50 mg/kg for 15 months. The difference between the performance of the 19 month-old treated with saline is statistically different (P $\leq$ 0.01) from the other two groups. 30-40 rats per group have been used.

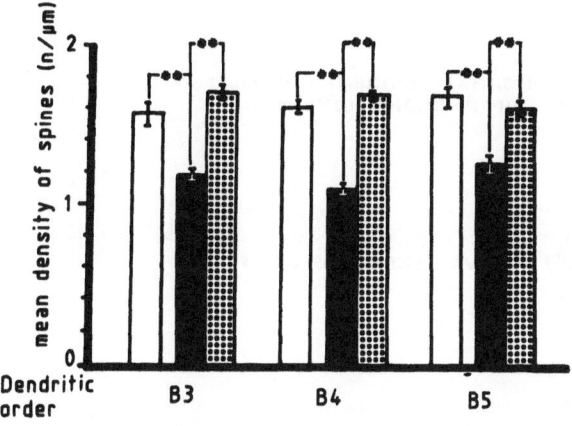

Fig. 4.

Spine density of pyramidal cell basal dendrites (of 3th, 4th, 5th order) in the CA1 sector of the hippocampus of young (4 months, white column) old (24 months, black column) and old BC-PS treated (dotted column) rats.
BC-PS has been given orally at a daily dose of 50 mg/kg for 29 months. Control group received saline.
** $P < 0.01$ (one way ANOVA).

In parallel, BC-PS prevents the age-induced loss of spines in the basal dendritic arborization of the CA_1 pyramidal cells of the dorsal hippocampus. This area is believed to play a major role in memory and learning function (Gray and McNaughton, 1983) and undergoes biochemical and morphological changes with age. The morphometric analysis of spine density, performed by a computerized image analysis system (IBAS) (Nunzi et al., 1986) indicates that prolonged oral BC-PS treatment is able to prevent the loss of dendritic spines in the hippocampal pyramidal cells, while maintaining the complexity of the dendritic arborization (Fig. 4); an effect which may account for the prevention of learning deficits.

CONCLUSION

We report here that BC-PS treatment is capable of (i) restoring age-dependent EEG abnormalities, (ii) improving memory deficit and preventing the decrease of learning capacity occurring in aged rats, (iii) preventing the age-dependent dendritic spine loss in the rat hippocampus. These data confirm previous results (Aporti et al., 1985; Drago et al., 1981; Corwin et al., 1985; Pedata et al., 1985) and are consistent with the finding that serine phospholipids show synergism with nerve growth factor (Bruni et al., 1982) and with compounds activating the protein kinase C (Battistella et al., 1985). Parallel experiments show that serine phospholipids together with the activation of cerebral function are also able to interact and activate the cells involved in the immune response (Bruni et al., 1986; Boarato et al., 1986).

To explain all these apparently unrelated observations we propose a role of serine phospolipids in the reactions of natural defense mechanisms. This role can be mimicked and even enhanced by the exogenous administration of phosphatidylserine and lysophosphatidylserine. The basic

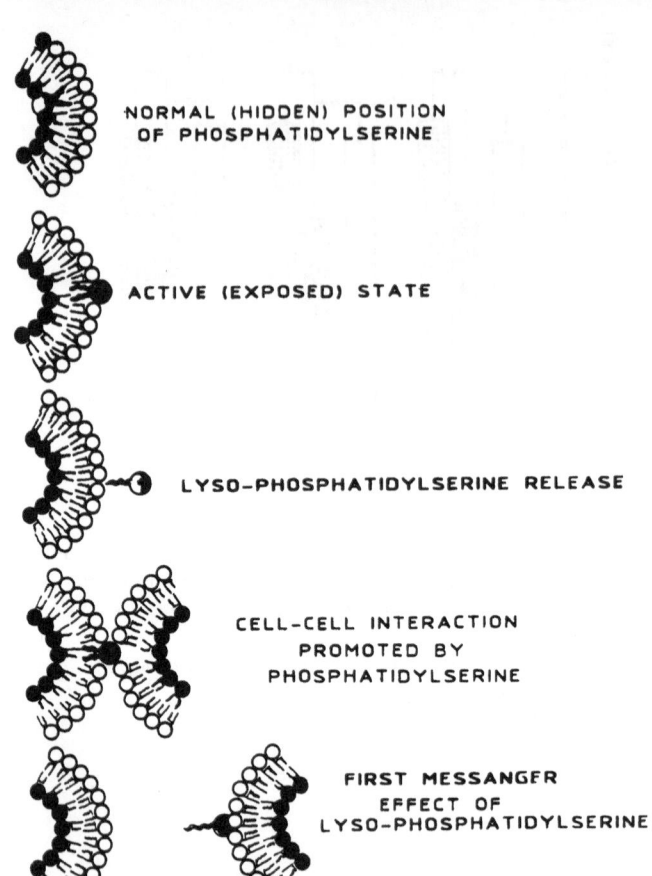

NORMAL (HIDDEN) POSITION
OF PHOSPHATIDYLSERINE

ACTIVE (EXPOSED) STATE

LYSO-PHOSPHATIDYLSERINE RELEASE

CELL-CELL INTERACTION
PROMOTED BY
PHOSPHATIDYLSERINE

FIRST MESSANGER
EFFECT OF
LYSO-PHOSPHATIDYLSERINE

Fig. 5.

Role of serine phospholipids in cell communication. Exposed phosphatidylserine may promote cell-cell interaction; alternatively the lyso-
derivative can be released to transmit a signal to target cells (first
messenger effect).

assumption of this proposal is the asymmetric phosphatidylserine distribution in the plasma membrane of undamaged eukaryotic cells (Verkleij et
al., 1973). Due to prevalent distribution in the inner leaflet of the
plasma membrane, phosphatidylserine does not interact with the external
cell environment under normal cell life (hidden position). Upon cell
damage, phosphatidylserine is exposed (active state) and may act as a
first messenger for those cells having a recognition system for this
phospholipid. Alternatively, lysophosphatidylserine is generated as a
soluble messenger, signalling that damage has occurred, and that repair
is needed (Fig. 5).

In order to explain the peculiar sensitivity of central nervous
system to the action of serine phospholipids, we are considering that the
small amount of these compounds crossing the blood-brain barrier may
produce a catalytic stimulation of membrane-linked events, beneficial to
neuronal activity. This possibility is supported by recent findings
showing that lysoPtdSer at micromolar concentrations, stimulates the rat
brain acyltransferase and therefore the incorporation of unsaturated
fatty acids into phosphatidylinositol and phosphatidylcholine
(Sbasching-Agler and Pullarkat, 1985). Since several membrane activities
such as receptor mobility, channel function and exocytosis are dependent
on the proper degree of acyl chain unsaturation, this effect may be

useful when renewal of the phospholipid fatty acids is required. Such an effect might be sufficient to explain the beneficial effect of prolonged PtdSer administrations on the activity of the aged brain as we have here reported. Alternatively, the effect of serine phospholipids on central nervous system _in vivo_ may result from the action of these compounds on peripheral tissues, chiefly the cells of the immune system. Communications between the brain and the immune system are well established (Besedovsky et al., 1983). Furthermore, mast cells and macrophages which are target cells for serine phospholipids in rodents, are known to produce and release trophic factors for several cells. The two possibilities are not mutually exclusive.

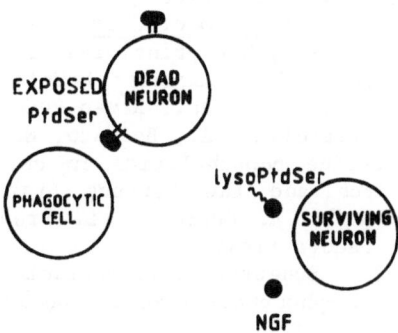

Fig. 6.

An attractive hypothesis explaining the simultaneous influence of serine phospholipids on the cells promoting defense reactions and on neurons is illustrated in Fig. 6. Dead neurons are in the suitable conditions to expose phosphatidylserine to the extracellular environment and to release lysophospatidylserine. These phospholipids may enhance the effect of the limited supplies of nerve growth factor and improve neuronal survival. In addition, the effect of serine phospholipids on phagocytic cells may help in the disposal of degenerating neurons. This hypothesis is supported by the observation of a synergism between serine phospholipids and nerve growth factor in mast cells, by the existence of recognition sites for phosphatidylserine in phagocytic cells, and by the large amount of this phospholipid in the nervous tissue. This mechanism may become operative any time in which accelerated neuronal death is occurring.

Hypothetical role of serine phospholipids in the central nervous tissue. A dead neuron exposes PtdSer which activates the macrophages and generates lysoPtdSer. The macrophage may ingest the degenerating neuron whereas lysoPtdSer may act synergistically with nerve growth factor (NGF) on a surviving neuron.

143

REFERENCES

Aporti, F., Borsato, R., Calderini, G., Rubini, R., Toffano, G., Zanotti, A., Valzelli, L., and Goldstein, L, 1985, Age-dependent spontaneous EEG bursts in rats: effects of brain phosphatidylserine, Neurobiol. Aging, 7:115.

Battistella, A., Mietto, L., Toffano, G., Palatini, P., Bigon, E., and Bruni, A., 1985, Synergism between lysophosphatidylserine and the phorbol ester tetradecanoylphorbolacetate in rat mast cells, Life Sci., 36:1581.

Besedovsky, H.O., Del Rey, A.E., and Sorkin, E., 1983, What do the immune system and the brain know about each other?, Immunology Today, 44:342.

Boarato, E., Mietto, L., Toffano, G., and Bruni A., 1986, Uptake of phosphatidylserine vesicles by rat leukocytes, Biochim. Biophys. Acta, submitted.

Bruni, A., Bigon, E., Boarato, E., Mietto, L., Leon, A., and Toffano, G., 1982, Interaction between nerve growth factor and lysophosphatidylserine on rat peritoneal mast cells, FEBS Letters, 138:190.

Bruni, A., and Palatini, P., 1982, Biological and Pharmacological Properties of Phospholipids, Progr. Med. Chem., 19:111.

Bruni, A., and Toffano, G., 1982, The principles of Phospholipid Pharmacology, in: "Transport in Biomembranes; Models Systems and Reconstitution," R. Antolini, et al., eds., Raven Press, New York.

Bruni, A., Mietto, L., Battistella, A., Boarato, E., Palatini, P., and Toffano, G., 1986, Serine phospholipids in cell communication, in: "Phospholipids Research and the Nervous System: Biochemical and Molecular Pharmacology," L.A. Horrocks, L. Freysz, and G. Toffano, eds., Liviana Press, Padova (Italy).

Calderini, G., Teolato, S., Bonetti, A.C., Battistella, A., and Toffano, G, 1981, Effect of lyso-phosphatidylserine on rat hypothalamic cAMP, in vivo, Life Sci., 28:2367.

Calderini, G., Bellini, F., Bonetti, A.C., Galbiati, E., Rubini, R., Zanotti, A., and Toffano, G., 1986, Pharmacological properties of phosphatidylserine in the aging brain: biochemical aspects and therapeutic potential, in: "Phospholipids Research and the Nervous System: Biochemical and Molecular Pharmacology," L.A. Horrocks, L. Freysz, and G. Toffano, eds., Liviana Press, Padova (Italy).

Casamenti, F., Mantovani, P., Amaducci, L., and Pepeu, G., 1979, Effect of phosphatidylserine on acetylcholine output from the cerebral cortex of the rat, J. Neurochem., 32:529.

Castagna, M., Takai, Y., Kaibuchi, K., Sano, K., Kikkawa, U., and Nishizuka, Y., 1982, Direct activation of calcium-activated, phospholipid-dependent protein kinase by tumor-promoting phorbol esters, J. Biol. Chem., 257:7847.

Corwin, J., Dean, R.L., Bartus, R.T., Rotrosen, J., and Watkins, D.L., 1985, Behavioral effects of phosphatidylserine in the aged Fischer 344 rat: amelioration of passive avoidance deficits without changes in psychomotor task performance, Neurobiol. Aging, 6:11.

Cunningham, C.M., Kingzette, M., Richards, R.L., Alving, C.R., Lint, T.F., and Gewurz, H., 1979, Activation of human complement by liposomes: a model for membrane activation of the alternative pathway, J. Immunol., 122:1237.

Drago, F., Canonico, P.L., and Scapagnini, U., 1981, Behavioral effects of phosphatidylserine in aged rats, Neurobiol. Aging, 2:209.

Franck, P.F.H., Chin, D.T.Y, Op den Kamp J.A.F., Lubin, B., Van Deenen, L.L.M., and Roelofsen, B, 1983, Accelerated transbilayer movement of phosphatidylcholine in sickled erythrocytes, J. Biol. Chem., 258:8436.

Gordesky, S.E., and Marinetti, G.V., 1973, The asymetric arrangement of phospholipids in the human erythrocyte membrane, Biochim. Biophys. Res. Commun., 50:1027.

Gray, J.A., and McNaughton, N., 1983, Comparison between the behavioural effects of septal and hippocampal lesions: a review, Neurosci. Biobehav. Reviews, 7:119.

Hudson, L.D.S., Fiddler, M.B., and Desnick, R.J., 1979, Enzyme therapy. X. Immune response induced by enzyme- and buffer-loaded liposomes in C3H/HeJ Gus^h mice, J. Pharmacol. Exp. Ther., 208:507.

Mazzari, S., and Battistella, A., 1980, Phosphatidylserine effects on dopamine release from striatum synaptosomes, in: "Multidisciplinary approach to brain development," C. Di Benedetta, R. Balàzs, G. Gombos, and G. Porcellati, eds., Elsevier, North-Holland.

Murphy, K.M.M., Gould, R.J., Oster-Granite, M.L., Gearhart, J.D., and Snyder, S.H., 1983, Phorbol ester receptors: autoradiographic identification in the developing rat, Science (New York), 222:1036.

Nishizuka, Y., 1984, The role of protein kinase C in cell surface signal transduction and tumor promotion, Nature (London), 308:693.

Nunzi, M.G., Milan, F., Guidolin, D., and Toffano, G., 1986, Dendritic spine loss in hippocampus of aged rats. Effects of brain phosphatidylserine administration, Neurobiol. Aging, Submitted.

Op dem Kamp, J.A., 1979, Lipid asymmetry in membranes, Ann. Rev. Biochem., 48:47.

Palatini, P., Dabbeni-Sala, F., and Bruni, A., 1972, Reactivation of a phospholipid-depleted sodium, potassium-stimulated ATPase, Biochim. Biophis. Acta, 288:413.

Papahadjopoulous, D., 1978, Calcium-induced phase changes and fusion in natural and model membranes, in: "Membrane Fusion," G. Poste, and G.L. Nicolson, eds, Elsevier, North Holland, Amsterdam.

Pedata, F., Giovannelli, L., Spignoli, G., Giovannini, M.G., and Pepeu, G., 1985, Phosphatidylserine increases acetylcholine release from cortical slices in aged rats, Neurobiol. Aging, 6:337.

Pepeu, G., Gori, G., and Bartolini, L., 1980, Pharmacologic and therapeutic perspectives on dementia: an experimental approach, in: "Aging of the brain and dementia", eds., L. Amaducci, A.N. Davison, and P. Antuono, eds., Raven Press, New York.

Richards, R.L., Gewurz, H., Siegel, J., and Alving, C.R., 1979, Interactions of C-reactive protein and complement with liposomes, II. Influence of membrane composition, J. Immunol., 122:1185.

Sbaschnig-Agler, M., and Pullarkat, R.K., 1985, Lysophosphatidylserine dependent incorparation of acylCoA into phospholipids in rat brain microsomes, Neurochem. Int., 7:295.

Sundler, R., Duzgunes, N., and Papahadjopoulos, D., 1981, Control of membrane fusion by phosholipid head groups. II. The role of phosphatidylethanolamine in mixtures with phosphatidate and phosphatidylinositol, Biochim. Biophys. Acta., 649:751.

Toffano, G., Leon, A., Mazzari, S., Savoini, G., Teolato, S., and Orlando, P., 1978, Modification of noradrenergic hypothalamic system in rat injected with phosphatidylserine liposomes, Life Sci., 23:1093.

Toffano, G., and Bruni, A., 1980, Pharmacological properties of phospholipid liposomes, Pharm. Res. Commun., 12:829.

Verkleij, A.J, Zwaal, R.F., Roelofsen, B., Comfurius P., Kastelijn D., and van Deenen, L.L.M., 1973, The asymmetric distribution of phospholipids in the human red cell membrane. A combined study using phospholipases and freeze-etch electron microscopy, Biochim. Biophys. Acta., 323:178.

Wu, P.S., Tin, G.W., and Baldeschwieler, J.D., 1981, Phagocytosis of carbohydrate-modified phospholipid vesicles by macrophage, Proc. Natl. Acad. Sci. USA, 78:2033.

Zanotti, A., Aporti, F., Toffano, G., and Valzelli, L., 1984, Effects of phosphatidylserine on avoidance relearning in rats (Preliminary observations), Pharm. Res. Commun., 16:485.

Zanotti, A., Valzelli, L., and Toffano, G., 1986, Partial reversal of scopolamine-induced amnesia by phosphatidylserine in rats, European J. Pharm., in press.

Zwall, R.F.A., 1978, Membrane and lipid environment in blood coagulation, Biochim. Biophysic. Acta, 515:163.

EFFECTS OF LECITHIN ON MEMORY AND LEARNING[1]

Hardo Sorgatz

Institute of Psychology
Technical University
of Darmstadt, D6100
Federal Republic of Germany

INTRODUCTION

The effects of cholinergic agonists and antagonists on mnestic performance has been documented by a variety of experimental studies with animals and human beings. A summary on this topic was published by Drachman & Sahakian in 1979. These studies seem to substantiate the theory of a performance impairing effect of cholinergic antagonists, whereas a hypothesis of a memory supporting function of cholinergic agonists currently lacks empirical support and is based only on theoretical assumptions. Mainly for patients with presenile and senile dementia it could be demonstrated that mnestic impairment can be decreased by choline, physostigmine and lecithin (Etienne et al., 1979; Vroulis et al., 1981; Tweedy & Garcia, 1982; Wood & Allison, 1982; Johns et al., 1983 and Thal et al., 1983). On the other hand, Christie et al. (1979) report that lecithin had no effect on the memory performances of 12 patients with Morbus Alzheimer and only for a very few patients a tendency toward improved performance was observable.

Studies on the effect of the cholinesterase-inhibitor physostigmine were conducted by Drachman & Leavitt (1974), Davis et al. (1978) and Sitaram et al. (1978) with normal adolescent subjects in which mostly positive results were obtained. It could also be shown that a slight overdose can lead to a reduction in performance. According to Drachman & Sahakian (1979) this is a consequence of the already existent optimal cholinergic level of functioning of the mostly adolescent subjects which was disturbed by additional choline. In the here presented study, however, the subjects exceeded an age of 44 years, which would suggest a suboptimal cholinergic level. Hence, a supportive effect of lecithin on the subject's mnestic functions was expected.

---

1) This paper is based on an article published in Fortschritte der Medizin, 1986, 104, 643-646.

METHOD

Subjects

78 persons were recruited by advertisements in newspapers. Alcohol dependent persons were excluded. The lecithin substance[2] was known to all subjects. None of them had consumed the lecithin containing substance within the last six months. 58 persons attended each of the three test sessions. As two of them did not complete all tests, the data of only 56 persons (47 female) could be examined.

The schooling of 45 subjects was comparable to British 'O' level standards, the rest of them had a preacademic level. The average age was 48 (SD = 4, Range = 44-56 years).

Tests

As a screening device the Benton-Test (Benton, 1945; Spreen, 1974) was used, which is especially sensitive to cerebral pathologies. In the multiple choice form it consists of 15 geometric configurations which have to be recognized out of four options after a short break. Two parallel forms (A and B) exist. As this is a purely clinical test and only measures low performance exactly it was applied in this study to single out subjects with cerebral damage. The test application lasts for five minutes.

The Learning and Memory Test (LGT-3; Bäumler, 1974) consists of six subtests (copying a town map, learning 20 Turkish words, giving the names to 20 objects shown in pictures, coordinating 13 telephone numbers (consisting of three units) with the right connection, memorizing 24 details of a building project, fitting the outlines of firm symbols to the right firm names). In the learning phase all items of the subtest had to be memorized and were then to be recalled directly after this test phase. The application of the test takes 35 minutes.

Due to the variety of test tasks multiple mnestic performances are required. Factor analytic studies have extracted the two factors of non-verbal and verbal memory as the most reliable dimensions. It must be noted, though, that the factor "non-verbal memory" is only defined by two subtests (town map, firm symbols), whereas, the remaining four subtests load high on "verbal memory". Apart from this, the subtest "town map" did not prove to be very reliable (r = .47) in this study. In both experimental groups an uninterpretable, significant drop in performance from pre to post was observed. Therefore the two subtests for non-verbal memory were not evaluated.

Two parallel forms (A and B) of the LGT-3 are available. As their correspondence is only moderate (parallel test reliability between r = .51 and r = .69) this was com-

---

2) Buerlecithin liquid/Manufacturer Roland Arzneimittel GmbH, D 2000 Hamburg 73

pensated by a counterbalanced order of tests applications
(Tab. 1).

Substances and Dosage

The verum substance contained in 100 ml nine grams of fluid
lecithin (a lipid complex extracted from plants with the effective components phosphatidylcholine, kephaline, inositphos-
phatit) and 16.4 ml alcohol. The original composition without
lecithin was applied as a placebo. As the alcohol content
was identical in verum and placebo, the effects of alcohol
on cerebral performance had not to be considered.

Tab. 1. Design

			Testform				Mean #
Session I	1st week	Benton-Test	A	B	A	B	
		LGT-3	A	B	A	B	M1
		Groups:	Lec.	Pla.			
		n =	16	13	15	12	
Session II	5th week	LGT-3	B	A	B	A	M2
Session III	9th week	LGT-3	A	B	A	B	M3
		LGT-3	B	A	B	A	M4

The two randomly composed groups consumed 30 ml of fluid
verum or placebo, respectively, every morning and 60 ml every
evening on each day between the test sessions. To control
compliance the subjects were asked to rate the taste of the
product every day on a visual analogue scale and to send in
the forms every weekend. Only subjects who fulfilled this
task on time were considered for further analyses.

Double-Blind Procedure

Neither the subjects nor the test conductors knew about
the hypothesis of the study or the application of either
placebo or verum. Respective questions were answered with
hints toward probably different taste qualities of the used
substances.

Experimental Procedure

The Benton-Test (five minutes) was applied at the beginning
of the baseline session to exclude subjects with possible
cerebral disturbances. Thereafter the LGT-3 (35 minutes) was
applied in one of its parallel forms. Half of the subjects
were supplied with form A, the other half with form B. Considering the applied test form, sex, and test results in
order to obtain comparable subsamples, the subjects were
split into a placebo and a verum group. As some subjects did
not complete all trials, parallelity of the two subsamples
could not be reached completely . The allocation of subjects
to treatments took place on the following day.

The alternative parallel test form of the LGT-3 (form B or A, respectively) was applied in the second test session which took place four weeks later at the same time of day (7:00 p.m.). After a further four weeks, both parallel test forms were applied with a break of five minutes. This third test session again took place at the same time. Table 1 gives a summary of the experimental plan. With respect to the age limits of the standardization sample the raw test values of the LGT-3 were not transformed into standard scores. As the distribution of test values showed no deviation from the normal distribution hypothesis t-tests and product-moment correlations were utilized as parametric tests.

RESULTS

To check the appropriateness of the LGT-3 for subjects in the age range from 44 to 56 years with average intelligence the mean values of the experimental sample (n = 56) were compared to those of the standardization sample (Tab. 2). As the test scores of the two experimental groups verum and placebo do not significantly differ with regard to the two parallel forms and to the four subtests the results are presented collectively.

Tab. 2. Ungrouped means for the Learning and Memory Test

Subtests	Norms	M1*	M2	M3	M4
Turkish	10.5	7.3	8.3	8.6	9.1
Building	11.0	5.6	7.2	9.1	8.1
Objects	9.5	7.9	8.7	9.9	9.4
Telephone	6.0	3.6	3.8	4.1	3.7

----

* see Tab. 1

An increase in performance from the first to the third test session is obvious for all four subtests on verbal memory. Highly significant results were obtained for all the subtests but "telephone-numbers". Significant improvements were also observable from the first to the third test session (M1 and M4) for both groups again with exception of "telephone-numbers".

A drop in performance in the subtest "telephone-numbers" was obtained for the placebo group (M1 = 3.9; M4 = 3.4; t = 0.94, df = 26, p = .36), whereas an insignificant increase in performance in the lecithin group was observed between the first and the fourth test (M1 = 3.3; M4 = 4.0; t = -1.42, df = 28, p = .17). Respective results were obtained for the comparison of the values of M1 and M3. The performance increase for the lecithin group in respect to the "telephone-numbers" was highly significant (M1 = 3.3; M3 = 4.2; t = -2.82, df = 28, p = .009). However, no changes were observable for the placebo group (t = -0.07, df = 26, p = .947).

The group values diverged most obviously in terms of the

choline hypothesis during the four week interval between the
first and second test session. For this period the increase
in memory performance of the lecithin group in all of the
four subtests for verbal memory were considerably higher
than those of the placebo group (Tab. 3). As the test re-
petition was expected to lead to a performance increase in
both groups and as a memory increase was also obtained by
authors of other studies the hypothesis is to be stated as
one sided, and hence a statistic one-tailed test was performed.

Tab. 3. Grouped Scores on LGT-3 for Session I and II

	PLACEBO				LECITHIN			
Subtest	M1*	M2	t**	p	M1	M2	t**	p
Turkish	7.1	7.5	-0.77	.22	7.6	9.0	-3.68	.00
Building	6.1	6.9	-1.44	.08	5.1	7.5	-3.07	.00
Objects	7.9	8.4	-0.89	.19	8.0	8.9	-1.65	.05
Telephone	4.1	4.2	-0.14	.45	3.3	3.8	-1.23	.11

* see Tab. 1
** t-test for correlated samples, one tailed

DISCUSSION

The comparison of the values of the LGT-3 which were
obtained in this study with those stated in the test manual
suggests that this Learning and Memory Test is also appli-
cable to person older than 44 years. The obtained mean values
are still lower than those of the standardization sample,
even after having been tested four times. This may be explained,
though, by the low standard of education of our subjects as
well as by their age dependant decrease on mnestic performances.

The results presented in Tab. 3 are based on the hypothesis
that choline agonists and hence also lecithin may improve
mnestic performance, a hypothesis that has been discussed
for some years. The positive effect of orally applied lecithin
on the synthesis of acethylcholine is unquestioned as well
as the importance of cholinergic synapses in the hippocampus
or limbic system, respectively, has been proved by a variety
of experimental studies. Nevertheless these increases in
mnestic functioning still lack an explicit explanation on
how a very unspecific manipulation of the choline metabolism
improved the mnestic capabilities of our subjects.

Drachman & Leavitt (1974) concluded from their results
that the age-dependent performance decrease of mnestic and
cognitive functioning may partly be due to a decrease of the
acetylcholine synthesis. This is also supported by post mortem
brain-analytic studies on elder and demented patients, where
an impairment of the cholinergic system but not of other
transmitter systems was observed. Furthermore, it could be
shown that the mnestic performance decrease which was induced
by choline antagonists could not be compensated by ampheta-
minetype substances which increase vigilance, whereas the

application of choline showed a positive effect. The impairment influenced more the process of acquisition than the consolidation and reproduction of the displayed material (Drachman & Sahakian, 1979; Wood & Allison, 1982; Johns et al., 1983).

In our experiment the interval between the acquisition and the reproduction of the LGT-3 items was only a few minutes long, so mainly short term memory performances were investigated. As no changes from the second to the third session were observed we may conclude that the effects on the choline metabolism did not improve long term memory performances but influenced mainly the acquisition of unknown information. If the assumed suboptimal cholinergic level of functioning for our 44 - 56 years old subjects was improved by the application of lecithin, this must have specifically effected the acquisition phase.

REFERENCES

Bäumler, G.   Lern- und Gedächtnistest LGT-3. Handanweisung. Göttingen: Hogrefe, 1974.
Benton, A.L.   A visual retention test for clinical use. Archives of Neurology and Psychiatry, 1945, 54, 212-216.
Christie, J.E., Blackburn, I.M., Glen, A.I.M., Zeisel, S., Shering, A. & Yates, C.M.   Effects of choline and lecithin on CSF choline and on cognitive function in patients with presenile dementia of the Alzheimer type. In: A. Barbeau, J.H. Growdon, and R.J. Wurtman (Eds.) Nutrition and the Brain. New York: Raven Press, 1979, 377-469.
Davis, K., Mohs, R., Tinklenberg, J., Pfefferbaum, A., Hollister, L. & Kopell, B.   Physostigmine: improvement of long-term memory processes in normal humans. Science, 1978, 201, 272-274.
Drachman, D. & Leavitt, J.   Human memory and the cholinergic system: a relationship to aging? Archives of Neurology, 1974, 30, 113-121.
Etienne, P., Gauthier, S., Dastoor, D., Collier, B. & Ratner, J.   Alzheimer's disease: Clinical effect of lecithin treatment. In: A. Barbeau, J.H. Growdon & R.J. Wurtman (Eds.)   Nutrition and the Brain. New York: Raven Press, 1979, 389-396.
Johns, C.A., Greenwald, B.S., Mohs, R.C. & Davis, K.L.   The cholinergic treatment strategy in aging and senile dementia, Psychopharmacology Bulletin, 1983, 19, 185-197.
Sitaram, N., Weingartner, H. & Gillin, J.   Human serial learning: Enhancement with arecholine and choline and impairment with scopolamine. Science, 1978, 201, 274-276.
Spreen, O.   Der Benton-Test. Handbuch. Bern: Huber, 1974.
Thal, L.J., Fuld, P.A., Masur, D.M., Sharpless, N.S. & Davies, P.   Oral physostigmine and lecithin improve memory in Alzheimer's disease, Psychopharmacology Bulletin, 1983, 9, 454-456.

Tweedy, J.R. & Garcia, C.A. Lecithin treatment of cognitively
     impaired Parkinson's patients. European Journal of
     Clinical Investigation, 1982, 12, 87-90.
Vroulis, G.A., Smith, R.C., Brinkman, S., Schoolar, J. & Gordon,
     J. The effects of lecithin on memory in patients with
     senile dementia of Alzheimer's type. Psychopharmacology
     Bulletin, 1981, 17, 127-128.
Wood, J.L. & Allison, R.G. Effects of consumption of choline
     and lecithin on neurological and cardiovascular
     systems, Federation Proceedings, 1982, 41, 3015-3021.

DISSEMINATION AND ACTIVITY OF AL 721 AFTER ORAL

ADMINISTRATION

Meir Shinitzky and Rachel Haimovitz

Department of Membrane Research
The Weizmann Institute of Science
Rehovot 76100, Israel

SUMMARY

Absorption of phospholipids in the small intestine is, in general, associated with hydrolysis by pancreatic lipases to water soluble products that eventualy reintegrate in the mucosa cells. An alternative route of phospholipid absorption is by endocytosis of small and tight assemblies like liposomes without hydrolysis. It might be therefore expected that shortly after absorption of phospholipids through one of these routes their composition in the blood stream is largely preserved. Along this rationale AL 721, a lipid mixture designated to reduce excess cholesterol from peripheral membranes, can be administered per os for in vivo application. A series of animal and human studies, where AL 721 was administered per os, lend support to this conclusion.

INTRODUCTION

Ample evidence is now available to support the empirical observation that consumption of plant phospholipids (e.g. commercial "lecithin") can help to reduce excess of serum cholesterol (Greten et al, 1980; Wilson et al, 1980; Murata et al, 1982; O'Mullane and Hawthrone, 1982). The mechanism of this effect presumably involves most of the metabolic pathways of lipids in the liver which determine the apparent level of serum lipoproteins and their lipid constituents: free cholesterol (C), esterified cholesterol (EC), phospholipids (PL) and triglycerides (TG). The most prominent effect on serum composition after chronic consumption of lecithin is a significant increase in mole ratio of high density lipoproteins (HDL) and low density lipoproteins (LDL). This increase is considered positive because it reflects a marked decrease in total serum cholesterol - an important determinant in disposition to atherosclerosis (Wilson et al, 1980).

Another positive aspect of the increase in HDL/LDL relates to the overall reduction in serum C/PL. This ratio is one of the most important regulatory factors of the level of free cholesterol in membranes, in particular the cell plasma membrane (Cooper and Strauss, 1984). Following this change in serum lipoproteins, increase in active uptake of HDL, where C/PL is around 0,3 M/M, and decrease in uptake of LDL, where C/PL is around 0,8 M/M, can have a marked effect on the total C/PL of cells (Goldstein and Brown, 1977). After metabolic processing, a decrease in C/PL of the cell plasma membrane is expected, especially when excess cholesterol resided initially in the membrane.

An independent route, the passive exchange, for the reduction in C/PL of cell plasma membranes upon increase in HDL/LDL operates by translocation of free cholesterol between pools of different C/PL levels (Cooper, 1977; Johnson et al, 1986). When the cell plasma membrane is the pool of the higher C/PL, efflux of cholesterol from the membrane to the serum will take place (Cooper, 1977; Johnson et al, 1986). The half-time of such a process is about several hours (Hagerman and Gould, 1951). In contrast to the active uptake of HDL or LDL this passive process is completely non-specific. It is, in principle, independent on the type of phospholipids in the serum and whether they are included in lipoproteins or in protein-free assemblies like liposomes for example. The passive process of changing the cholesterol level in cell plasma membranes operates well in vitro, even on isolated membranes (Cooper, 1977; Cooper and Strauss, 1984).

Inasmuch as the active process of reduction in membrane cholesterol which was described above involves the progressive increase in HDL/LDL, it is slow (weeks or months), in terms of reaching a lower stable C/PL level in the cell plasma membrane. The passive process with non-proteinic phospholipid assemblies is, in principle, faster. Yet, for practical application it possesses the inherent problem of how to introduce such assemblies into the blood stream for sufficiently long time to enable them to affect the level of membrane cholesterol.

Absorption and dissemination of lipids

The effect of specific lipids on cell plasma membrane composition and function, can be readily studies in tissue culture (Shinitzky, 1984). Such studies could provide the basis for subsequent in vivo application, especially in attempts to rectify well defined aberrations in cell membranes. The desired lipid can be introduced into the blood stream either per os or by intravenous injection or by an intermediate route such as intraperitoneal administration. The limiting factors in these routes are the metabolic alterations in the lipid composition before reaching the blood stream and the metabolic clearance by the liver and other lipid consuming tissues. These aspects are briefly overviewed in the following.

Lipid absorption through the small intestine. Over 90% of the lipid in animal and plant food is triglycerides. The process of lipid absorption in the digestive tract, which takes place almost exclusively in the small intestine, is therefore programmed for digestion and absorption of triglyceride rich lipid particles. The first step in the absorption of dietary lipids is conversion of the water-immiscible lipid particles into a fine and absorbable lipid dispersion. Two major processes contribute to this step. The first is enzymic hydrolysis of part of the triglycerides by pancreatic lipases, the products of which are free fatty acids and 2-monoglycerides. In the case of phospholipid rich diet, phospholipase A_2 digestion is added to produce lysophospholipids and free fatty acids. The second process is hormone stimulated secretion of phospholipids and bile salts from the gallbladder into the lumen of the small intestine. The products of all these steps cooperate to disperse and homogenize the remaining intact triglycerides which can then be absorbed by the mucosa cells on the intestinal lining (for recent reviews see Carey et al, 1983; Tso, 1985).

The processed lipid particles are absorbed by two independent passive processes. Single amphipathic molecules, but not intact triglyceride molecules, can partition into the ruffled membrane of the mucosa cells and from there to further partition into the cytoplasm. Alternatively, endocytosis of whole units of dispersed lipids can take place directly into the cytoplasm of the mucosa cells. It is not yet clear what are the factors which determine the mode of absorption of processed lipid in the lumen. From the scarced available data it seems that the looser is the physical bonding in the processed lipid dispersion the lesser is the chance for absorption by endocytosis (Carey et al, 1983). Phospholipids in the condensed liposomal assembly, for example, when introduced into the small intestine are taken up almost exclusively as intact particles by endocytosis (Patel et al, 1985). This issue, however, is of secondary importance since the subsequent passage of lipids from the mucosa cells to the blood via the lymph is the step which will affect, at least temporarily, the apparent lipid composition of the serum.

The degradation products of the triglycerides and phospholipids, i.e. fatty acids, 2-monoglycerides and lysophospholipids, can act as detergents above their critical micellar concentration. Therefore, an effective resynthesis of trigliceries and phospholipids in the cytoplasm of the mucosa cells takes place so that the lipids accumulated there are by and large of a similar pattern to those that were absorbed. In addition to that, in these cells apolipoproteins, in particular of the B class, are synthesize and integrated with the absorbed lipids to form assemblies which resemble serum chylomicrons and low density lipoproteins. These are then transported to the lymph and tunneled into the blood stream in the thoracic duct. The rate of clearance and processing by the liver of the newly absorbed lipids is assumed to be in the range of minutes to hours. This important parameter indicates that newly absorbed lipids can circulate in the blood stream for at least several cycles before being metabolized. In a phospholipid rich diet processing in the lumen and further in the mucosa cells is assisted mostly by pancreatic phospholipase A_2 and presumably by phospholipase C as well (Parthasarathy et al, 1974). Most of the degradation products (i.e. lysophospholipids, diglycerides, fatty acids and phosphoryl esters) probably to reassemble before secretion into the blood. On the whole, it seems that shortly after introduction into the blood the digested triglycerides or phospholipids retain most of their initial composition except that the acyl chains are reshuffled.

In conclusion, the arguments presented above seem to indicate that it is possible to design a diet of phospholipids and glycerides which will appear for a limited period of time in the blood as specific assemblies. Such assemblies may be designed for a specific activity such as for dissolving arterial lipid plaques or for extraction of excess membrane cholesterol. The microscopic composition of such assemblies initially differs from the bulk lipoproteins and chylomicrons but with time, these assemblies will assimilate with the rest of the serum lipids through passive diffusion exchange and metabolic processing by the liver. According to this hypothetical trend the relative content of such specialized lipid assemblies may be low and even undetectable by conventional analytical tools. They could be best monitored indirectly, e.g. through their activity.

Intravenous introduction of lipids. Intravenous alimentation with lipid emulsions is currently a routine practice in artificial feeding (Shenkin and Wretlind, 1978). The most widely used commercial preparation is Inralipid (Vitrum, Stockholm) which is composed of 10 or 20% lipid in aqueous 2.5% glycerol (which is equivalent to saline). The lipid in Intralipid is a mixture of 88% soybean triglycerides and 12% egg lecithin (in the 10% emulsion) or 94% soybean triglycerides and 6% egg lecithin (in the 20% emulsion). The average particle size in these emulsions is around 0.15 μm - similar to that of serum chylomicrons (Hallberg, 1965; Rossner, 1974). This emulsion is well tolerated both by animals and man up to levels of as high as

10gr/Kg/day. It is being utilized metabolically similar to post-meal chylomicrons at a rate of several hours (Carlson and Hallberg, 1963).

Intraperitoneal administration of lipids. The peritoneal cavity can provide a convenient store for slow release of lipids into the blood stream. In practice, such a route of administration is relevant to small experimental animals like mice or rats. Although there is not yet data available on the rate of diffusion of lipids from the peritoneum to the blood, based on similar experiments (Torres et al, 1978) it might be assumed that it is in the range of hours. Accordingly, in comparison with intravenous injection the effect expected from intraperitoneal administration of lipids might be similar to that obtained by slow intravenous infusion.

ABSORPTION AND DISSEMINATION OF AL 721

     AL 721 is a special mixture of lipids from hen egg yolk designed for extraction of excess cholesterol from cell plasma membranes (Lyte and Shinitzky, 1985), e.g. in aging (Shinitzky et al, 1983; Rabinowich et al, 1987) or drug addiction (Heron et al, 1982). It consists of 70% triglycerides, 20% phosphatidylcholine (PC) and 10% phosphatidylethanolamine (PE). The following parmacokinetic study was carried out by oral administration of AL 721 traced with radioactively labelled PC.

     1-10μCi PC labelled at the β-chain with ^{3}H-linoleic acid and/or at the headgroup with ^{32}P was introduced into 5mg AL 721 and used as one feeding dose. Each dose was dispersed in 1 ml distilled water and then introduced by force-feeding administration into the esophagus of a 2-3 months old male mouse (C57Bl) in the morning after 3 hours fast. The mouth and esophagus were then rinsed thoroughly with water to ensure the complete entrance of the lipid into the stomach. The animal was then allowed free access to food and water. The absorption and distribution of labelled PC were followed in groups of 4 animals per each post-ingenstion time.

     The average percent radioactivity retrieved in lipid extracts of the body at 1,3,5, 12 and 24 hours after feeding is presented in Figure 1. The highest level of radioactive PC in the body was found after 3 hours. The ratio of ^{3}H/^{32}P in this experiment remained close to 1 indicating that the absorbed radioactivity was retained in PC or other phospholipids. Part of the remaining PC was presumably metabolized to water soluble products (e.g. ^{32}PO$_4^{-3}$, ^{3}H$_2$O) or excreted. Most of the absorbed lipid was found to remain in the upper digestive tract (stomach and small intestine) for up to 12 hours as shown in Figure 2. The amount of PC retrieved in the blood, liver and in particular the brain was relatively low (Figure 3). It seems that the passage of AL 721 from the stomach to the small intestine (Figure 2) and from there to the blood (Figure 3) is unexpectedly slow. It is not yet clear whether the observed slow lipid absorption and transport are due to the regimen used or are special traits of lipid digestion in the mouse.

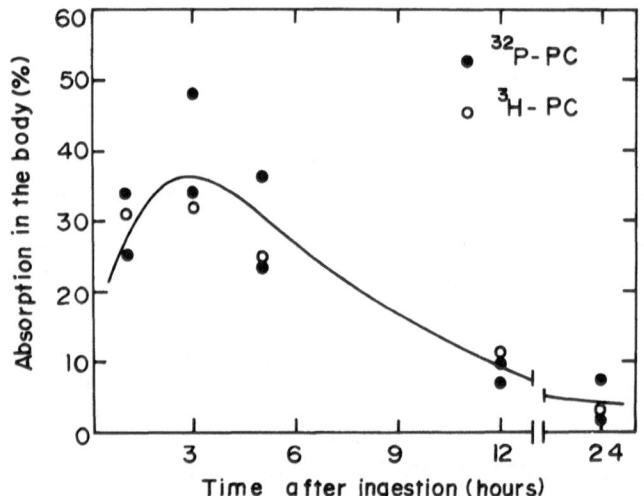

Fig. 1. Percent absorption of AL 721 by mice after oral administration. Total radioactivity counts were summed over relevant body organs (stomach, small intestine, large intestine, liver, spleen and brain) after ingestion of AL 721 containing ^{32}P-PC or ^{3}H-PC (5 mg, 1-10 μci per animal). Each time point represents the mean obtained for 4 mice.

Experiments with injected radioactive AL 721, analogous to the above, are at the moment inconclusive because of technical difficulties. We have nevertheless accumulated some information as to the effect of intravenous administration of AL 721 on lymphocyte activity (Mark Lyte, Ph.D. Thesis).

Four old mice (22-24 months C57B1) received a single
intravenous injection of 5mg AL 721 in 0.1ml saline. After
24 hours splenocytes were isolated and their mitogenic
stimulation with different doses of concanavalin A (ConA)
was determined by incorporation of ^{3}H-thymidine.
Splenocytes from 4 untreated old mice (22-24 months, C57B1)
served as control. The results, which are summarized in
Figure 4, indicate a several fold increase in lymphocyte

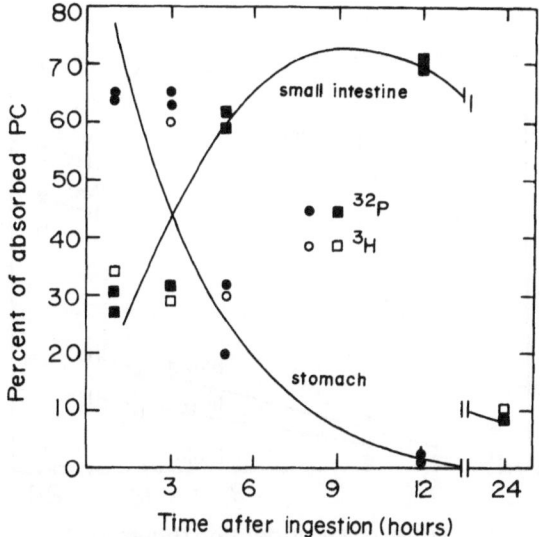

Fig. 2. Percent of absorbed PC in AL 721, traced by ^{32}P-PC
         or ^{3}H-PC (see legand to Figure 1), which resided in
         the stomach or the small intestine at different
         times after oral administration. Each time point
         represents the mean obtained with 4 mice.

responsiveness and partial abrogation of the suppressive
effect of high ConA doses in splenocytes from the AL 721
treated animals.

Augmentation of lymphocyte responsiveness in the aged by
AL 721 diet.

Studies with old mice under AL 721 diet indicated a
marked rectification of lymphocyte responsiveness (Shinitzky

et al, 1983; Shinitzky, 1986). Inasmuch as AL 721 is a par
excellence nutrient, no adverse reactions could be expected
when given in the diet to men. This was verified by acute
and chronic toxicology studies in animals, which opened the
way for studies with human subjects under a well defined
AL 721 diet. In our immunological study (Rabinowich et al,
1987) we have selected 10 participants over 75 years of age
who were immune suppressed but did not display any organic
disease and were not taking immunosuppressive drugs. The

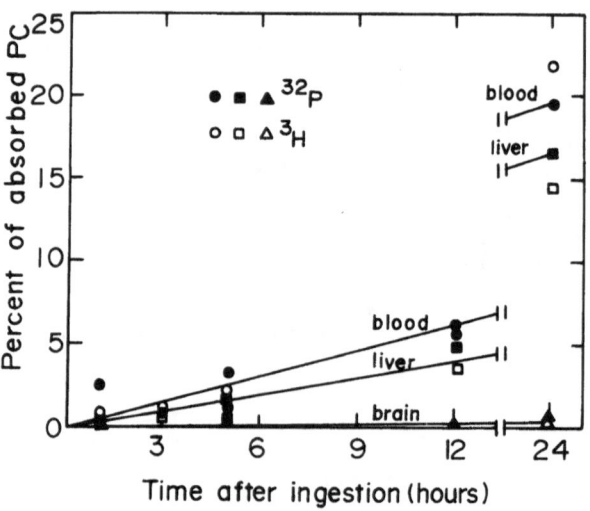

Fig. 3. Percent of absorbed PC in AL 721, traced by ^{32}P-PC
or ^{3}H-PC (see legens to Figures 1 and 2), retrieved
in the blood liver and brain at different times
after oral administration. Each time point
represents the mean obtained with 4 mice.

study and the experimental protocol were carried out at the
Meir Hospital, Kfar Saba, Israel and was conducted under the
supervision of Prof. A. Klajman and Dr. H. Rabinowich. The
mitogenic responsiveness of peripheral blood lymphocytes was
tested in each of the 10 participants at least 3 times
within 3 weeks before entering the study in order to assess
their basal immune competence. AL 721 diet of 10gr was
given each morning during a period of several weeks and the

responsiveness of peripheral blood lymphocytes to mitogens
was tested every 4-7 days. The AL 721 diet was then stopped
and the effect on mitogenic stimulation was measured 7 days
later. The following trend was observed: Already after
several days of AL 721 diet a significant increase in
response to mitogens was noticed in most participants.
After about 3 weeks it reached its highest level which in
some cases approached that of the young. Upon cessation of
the diet the lymphocyte responsiveness slowly declined
toward the initial basal level. The results obtained with
ConA stimulation are summarized in Figure 5.

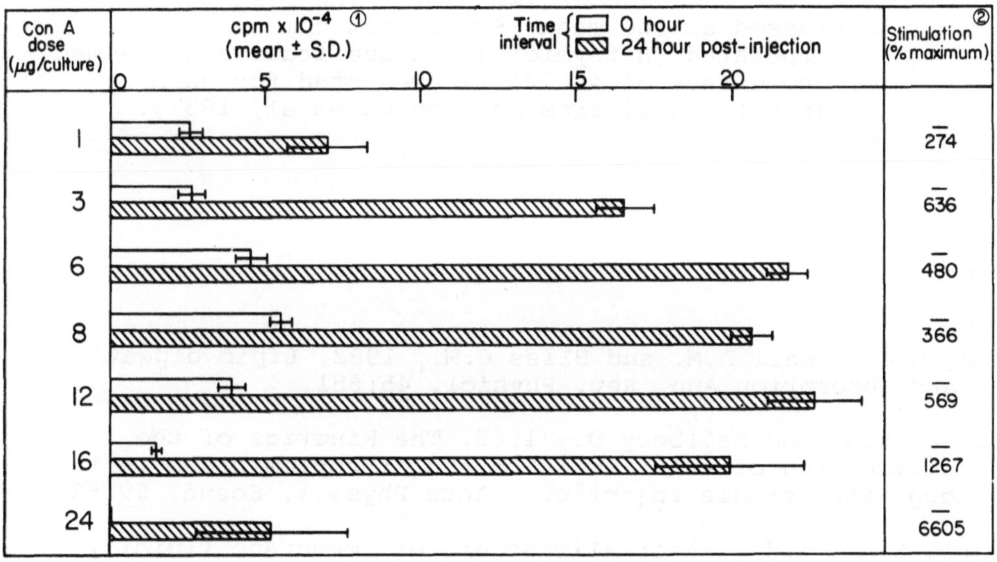

Con A dose (μg/culture)	cpm x 10^{-4} ① (mean ± S.D.)	Time interval {☐ 0 hour / ▨ 24 hour post-injection}	Stimulation ② (% maximum)
1			274
3			636
6			480
8			366
12			569
16			1267
24			6605

Fig. 4. Mitogenic stimulation of splenocytes from old mice
(22-24 months) by different doses of ConA 24 hours
after intravenous injection of 5 mg AL 721 per
mouse. The results are presented as mean +SEM of
incorporated ^{3}H-thymidine in splenocytes of 4
treated mice (▨) in comparison with splenocytes of
4 untreated mice (☐) of the same age group.

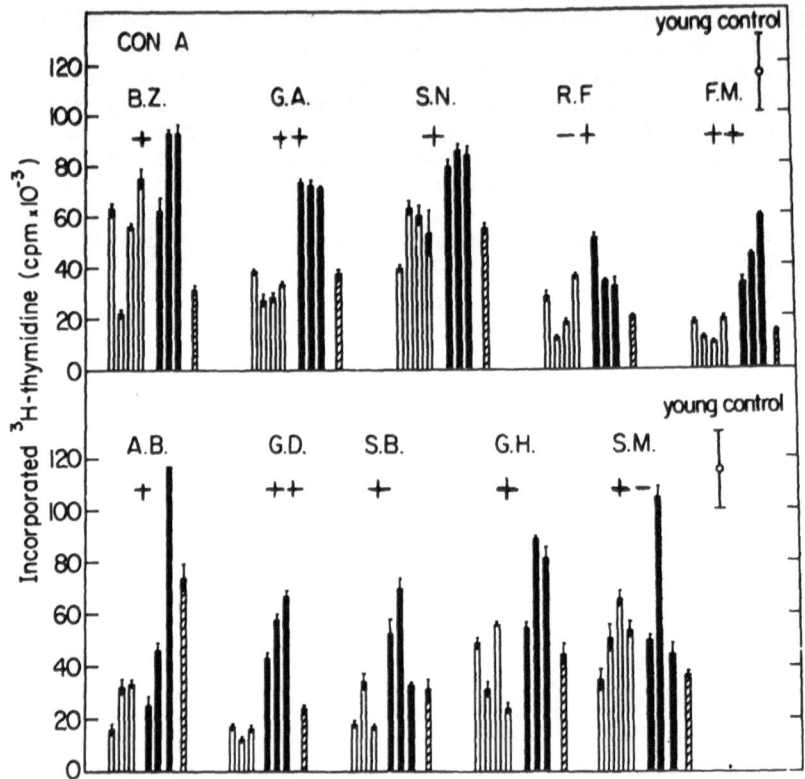

Fig. 5. Mitogenic stimulation of peripheral blood
lymphocytes from aged healthy subjects by ConA.
Lymphocytes were drawn and tested at weekly
intervals before (□), during (■) and after (▨)
the course of the AL 721 diet. The results are
expressed as the mean +SEM of net cpm of
incorporated ³H-thymidine. A qualitative assessment
of the effect of AL 721 is presented for each
subject (adopted from Rabinowich et al, 1987).

REFERENCES

Carey M.C., Small D.M. and Bliss C.M., 1983, Lipid digestion
and absorption Ann. Rev. Physiol. 45:651.

Carlson L.A. and Hallberg D., 1963, The Kinetics of the
elimination of a lat emulsion and of chylomicrons in the
dog after single injection. Acta Physiol. Scand. 59:52.

Cooper R.A., 1977, Abnormalities of cell membrane fluidity
in the pathogenesis of disease, N. Eng. J. Med. 197:371.

Cooper R.A. and Strauss J.F., 1984, Regulation of cell
membrane cholesterol, in physiology of membrane fluidity
(M. Shinitzky, ed) Vol. 1 pp. 73-98. CRC press, Boca
Raton, Florida

Goldstein J.L. and Brown M.S., 1977, The low-density
    lipoprotein pathway and its relation to atherosclerosis,
    Ann. Rev. Biochem. 46:897.

Greten J., Raetzer H., Stiehl A. and Schettler, 1980, The
    effect of polyunsaturated phosphatidylcholine on plasma
    lipids and fecal sterol excretion. Atherosclerosis
    36:81.

Hagerman J.S. and Gould R.G., 1951, The in vitro interchange
    of cholesterol between plasma and red cells, Proc. Soc.
    Exp. Biol. Med. 78:329.

Hallberg D., 1965, studies on the elimination of exogenous
    lipids from the blood stream. Acta Physiol. Scand.
    64:306.

Heron D., Shinitzky M. and Samuel D., 1982, Alleviation of
    drug withdrawal sympthons by treatment with a potent
    mixture of natural lipids. Eur. J. Pharmacol. 83:253.

Johnson W.J., Bamberger M.J., Latta R.A., Rapp P. E.,
    Phillips M.C. and Rothblat G.H., 1986, The bidirectional
    flux of cholesterol between cells and lipoproteins. J.
    Biol. Chem. 261:5766.

Lyte M. and Shinitzky M., 1985, A special lipid mixture for
    membrane fluidization. Biochim. Biophys. Acta 812:132.

Murata M., Imaizumi K. and Sugano M., 1982, Effect of
    dietary phospholipids and their constituent bases on
    serum lipids and apolipoproteins in rats J. Nutr.
    112:1805.

O'Mullane J.E. and Hawthrone J.N., 1982, A coparison of the
    effect of feeding linoleic acid-rich lecithin or corn
    oil on cholesterol absorption and metabolism in the rat.
    Atherosclerosis 45:81.

Parthasarathy S., Subbaigh P.V. and Ganguly J., 1974, The
    mechanism of intestinal absorption of
    phosphatidlylcholine in rats. Biochim. J. 140:503.

Patel H.M., Tuzel N.S. and Stevenson R.W., 1985,
    Intracellular digestion of saturated and unsaturated
    phospholipid liposomes by mucosal cells. Biochem.
    Biophys. Acta 839:40.

Rabinowich H., Lyte M., Steiner A, Klajman A. and Shinitzky
    M., 1987, Partial restoration of immune competence in
    aged humans under a special lipid diet (AL 721). Mech.
    Age. Dev. in press.

Rose H.G. and Oklander M., 1965, Analysis of erythrocyte
    membrane lipids J. Lipid Res. 6:428.

Rossner S., 1974, studies on intravenous fat tolerance rest.
    Methodological, experimental and clinical experiences
    with Intralipid. Acta Med. Scand. Suppl. 564:1.

Shenkin A. and Wretlind A., 1978, parenteral nutrition, Wed. Rev. Nutr. Diet. <u>28</u>:1.

Shinitzky M., Lyte M., Heron D. and Samuel D., 1983, Intervention in membrane aging - The development and application of Active Lipid, in Intervention in the aging process, part B. pp. 175-186. Alan R. Liss, Inc.

Shinitzky M., 1984, Membrane fluidity and cellular functions, in Physiology of membrane fluidity (M. Shinitzky, ed) Vol. 1 pp. 1-51.

Shinitzky M., 1986, The lipid regimen, in Alzheimer's and Parkinson's diseases (A. Fisher, I. Hanin and C. Lachman, eds) PR 593-602. Plenum Press, New York.

Torres I.J., Litterst C.L. and Guazino A.M., 1978, Transport of model compounds across the peritoneal membrane in the rat. Pharmacology <u>17</u>:330.

Tso P., 1985, Gastrointestinal digestion and absorption of lipid, Adv. Lipid Res. <u>21</u>:143.

Wilson P. W., Garrison R.J., Castelli W.P., Feinleib M., McNamara P.M. and Kannel W.B., 1980, Prevalence of coronary heart disease in the Framingham offspring study: role of lipoprotein cholesterols. Am. J. Card. <u>46</u>:649.

EFFECT OF DIETARY LECITHIN AND NATURE OF DIETARY FAT UPON GROWTH AND BILE COMPOSITION IN THE GERBIL

Tom R. Watkins and Anthony Pagano

Nutrition and Food Science, Hunter College
City University of New York
New York, NY USA

The American diet is rich in fat, particularly saturated fat. In this century the consumption pattern has shifted such that 40 - 45% of calories are provided by fat. In contrast, our ancestors at the turn of the century may have consumed 25 - 30% of their calories as fat. A rise in the incidence of cardiovascular disease, and other conditions of impaired fat digestion, absorption and transport has accompanied this dietary trend(1). Professional societies, e. g., the American Heart Association, and the Government have issued diet recommendations, urging Americans to reduce their fat intake to 30% of calories, and include unsaturated fats in their menus. Though no absolute proof exists linking the disease incidence with dietary practice, sufficient evidence exists that these groups have sounded the alarm calling for a change in eating habits for improved quality of life.

Ironically, during this period innovations in food, specifically fat, processing became widely practiced which may have exacerbated the health problems. A potential food oil would be routinely separated into the fractions: triglyceride; fatty acid; gum; and, lecithin. The consumer would purchase the purified, denuded vegetable oil, and consume generous amounts without the natural emulsifier, i. e., lecithin. Both visible and invisible fats would be consumed without the lecithin. That oil and water are immiscible has been observed universally. Yet we have continued to introduce generous amounts of fat into the body, a system comprised of 60% water, perhaps without considering the consequences. Have we neared a limit of relatively safe intake of dietary fat, especially fat without any natural emulsifier, compatible with health?

Several dietary states may perturb liver function: inadequate protein, excess alcohol, and starvation, among others. Impaired liver function may lead to poor bile quality, and subsequent biliary disease. Sixteen million Americans suffer with cholestatic bile, associated digestive problems, and often secondary malnutrition. Symptoms may include dyspepsia and severe abdominal pain after a fat-rich meal, cholestatic bile, and eventually the need for therapy. Therapy might be sonic treatment to disintegrate stones, gall bladder removal, or, perhaps chenodeoxycholate treatment, initially (2). No efficacious dietary treatment has been discovered or adopted, perhaps partly a result of late diagnosis.

Cystic fibrosis symptoms share common features with cholestasis. The cystic fibrotic patient (1/2400 live births among Caucasians) endures the consequences of poor quality bile. Though until recently most lived but a

few years and starved of malnutrition, unable to absorb adequate fat energy
to support growth and resilience against infection, one can now be expected
to live into the third decade, if special fat supplements and digestive
enzymes be taken daily. No cure is known, since the genetic defect(s) has
not been precisely identified. Exocrine pancreatic insufficiency with subsequent malabsorption, reduced energy availability and growth impairment,
poor bile quality with excessive stool losses of bile salts, copious mucous
secretion, excessive sweat loss of sodium, respiratory distress, skin, blood,
and adipose tissue anomalies, and signs of essential fatty deficiency
(reduced tissue linoleate) characterize the disease (3). Because of impaired
fat emulsification and absorption, the child usually falls in lower growth
chart velocities, often below the fifth percentile. To enhance availability
of dietary calories, pancreatic enzyme replacement therapy will be initiated
and followed throughout life, at least until past puberty and the growth
spurt. The serum essential fatty acid abnormalities and other symptoms may
never became normal; growth may begin to approach normal with ideal compliance and no severe respiratory infections. With poor fat digestion and
absorption, the cystic fibrotic child or young adult will lose much dietary
energy in the stool (steatorrhea) as well as bile salts, and fat-soluble
vitamins (4). The loss of bile salts in the fatty stools perturbs the
enterohepatic circulation of bile, reducing the pool size; composition of
bile and bile conjugates changes (5). Could a modified fat diet be devised
that would improve fat absorption by enhancing the emulsifying potency of
gall bladder bile produced by the liver cell? Can the diet modulate the.
function of the liver?

        Evidence exists that an altered dietary fatty acid pattern will change
liver and peripheral cholesterol metabolism, both in humans and animals.
Hegsted and McGandy (6), and many others later, have showed that altered
cholesterol levels in the diet, added as egg yolk, resulted in corresponding
modest changes in serum cholesterol. The serum cholesterol changed 5 mg/dl
for each 100 mg of dietary cholesterol added, a finely correlated mass
effect. A substitution of fatty acid in the diet exerted a marked effect
upon serum cholesterol. After feeding a saturated oil, such as coconut,
and shifting to the monounsaturate olive oil, and thence to the polyunsatur-
ate safflower oil, they observed a 40 mg/dl decrement associated with each
such shift. Indeed, the modulating influence of dietary fat upon cellular
metabolism of cholesterol, a chief bile constituent, has been demonstrated
in well controlled studies.

        On the other hand, manipulation of dietary lecithin levels has also
been demonstrated to influence liver and peripheral cholesterol metabolism.
Wong and her coworkers (7) have showed that even with the dietary fatty
acid composition held constant, tissue cholesterol metabolism responds to
the presence of dietary lecithin addendum. In the rhesus monkey, the inclusion of lecithin resulted in a 23% decrease in serum cholesterol, and
36% decrease in serum LDL cholesterol.   Animals fed a specific ration in
general comply better than free-living humans eating self-selected diets.

Purpose

        The purpose of the present study was to provide information about
two questions. First, will omission of lecithin from the diet alter hepatic
cell function, such that gall bladder bile salt composition be altered in
quantity or quality? Second, will the nature of dietary fat, whether saturated, polyunsaturated, or monounsaturated, modulate liver cell capacity to
provide bile with a normal profile of bile acid conjugates? On the basis
of data cited above, such dietary modulation would be expected, since the
bile salts contain the cholesterol nucleus; and, they are metabolized by
membrane-associated enzymes, which would be expected to respond to changes
in membrane chemical composition with respect to lecithin and/or fatty acid.

168

Diet & Methods

The animal used in the present studies was the Mongolian gerbil, Meri-
ones unguiculata. Weanling female animals, 30 grams average weight (Tumble-
brook Farms, Brookfield, MA), were housed communally in stainless steel mesh
cages. They were maintained in an air-conditioned room at 26° C. with a
twelve-hour photo period, 7 AM to 7 PM. The had access to drinking water
and food ad libitum. After receipt at the laboratory, animals were allowed
to adapt to the environs for two days before eating the defined diets.
Diet composition appears in Table 1. Groups contained 5 - 7 animals each.

The diet was chosen to mimic the current American diet which provides
more than 40% of calories as fat. Fat was provided as coconut oil, safflower
oil, or, in one resupplemented (RESUPP) group, peanut oil. Each diet
contained 20% total fat by weight, including 1.5% of safflower to prevent
the symptoms of essential fatty acid deficiency. The coconut oil diet with
with highly saturated fat (53% 12:0 and 14:0) has been associated with
atherogenic lesions. A generous level of safflower oil in the diet, on the
other hand, has been associated with reduced lesion formation, and reduced
risk of cardiovascular disease; safflower contains mainly polyunsaturated
acid (75% 18:2). Peanut oil contains an abundant amount of monounsaturated
acid (49% 18:1), as well as some polyunsaturated and saturated acid. Feeding
such a range of acids to the animals, with or without lecithin addendum,
possible influences of each constituent could thereby be tested with respect
to growth, liver and gut weight, and bile quality.

Animals were fed 7 to 10 days, fasted for 12 hours, and gassed with
CO_2. The liver and small bowel were excised, weighed, and gall bladder bile
aspirated. Tissue was stored at -20° C. until analysis. Bile salts were
analyzed by thin layer chromatography with fluorescence detection (366 nm
excitation; 500 nm bandpass), by the method of Levin & Touchstone [8].
Bile samples were diluted 20 to 100-fold before analysis with water:ethanol
(75/25). Scanning was done with a Shimadzu 910 with a CR1B integrator.
Reference standards were run at 100 to 1600 ng for calibration (Sigma, St.
Louis, MO).

Table 1.   Diet Formulae

Oil:	Coconut		Safflower		Peanut
Lecithin:	-	+	-	+	+
Script:	C	CL;CLR	S	SL;SLR	PLR
Ingredient, g.					
Casein[*]	15	15	15	15	15
Starch, corn	55	55	55	55	55
Cellulose	6.3	6.3	6.3	6.3	6.3
Salt mix[§]	6.3	6.3	6.3	6.3	6.3
Vitamin mix[§]	0.63	0.63	0.63	0.63	0.63
Fat					
Coconut	18.8	15.8	----	----	----
Safflower	1.2	1.2	20.0	17.0	----
Peanut	----	----	----	----	17.0
Lecithin[¶]	----	3.0	----	3.0	3.0

[*] Ingredients obtained from U. S. Biochemicals, Cleveland, OH
[§] Hegsted, D. M., et al. J. Nutr. 104: 588 (1974)
[¶] Kindly provided by Lucas Meyer, Decatur, IL

Results

Diet supplemented with lecithin supported better growth, regardless of the type of dietary lipid fed. The growth results of feeding the weanling gerbil these diets were recorded for 10 days after which the animals were sacrificed. Lecithin-deprived animals were re-supplemented (R) with lecithin for seven (7) days, and sacrificed on day 17. Groups previously fed coconut (C) or safflower oil (S) were re-fed coconut, safflower, or peanut (P) oil. Data appear in Figure 1. Though the safflower-fed animals given lecithin (SL) gained an average 3.0 g, the coconut oil group (CL) lost 1.5 g in the first period. If the animals were deprived of lecithin in either lipid group (C or S), they lost considerable weight, 7.5 g in the coconut oil group (C), and 5.5 g in the safflower oil group (S). During the lecithin re-supplementation period the mean weight gains were: coconut (CLR), 1.8 g; safflower (SLR), 6.1 g; and peanut oil (PLR), 5.8 g.

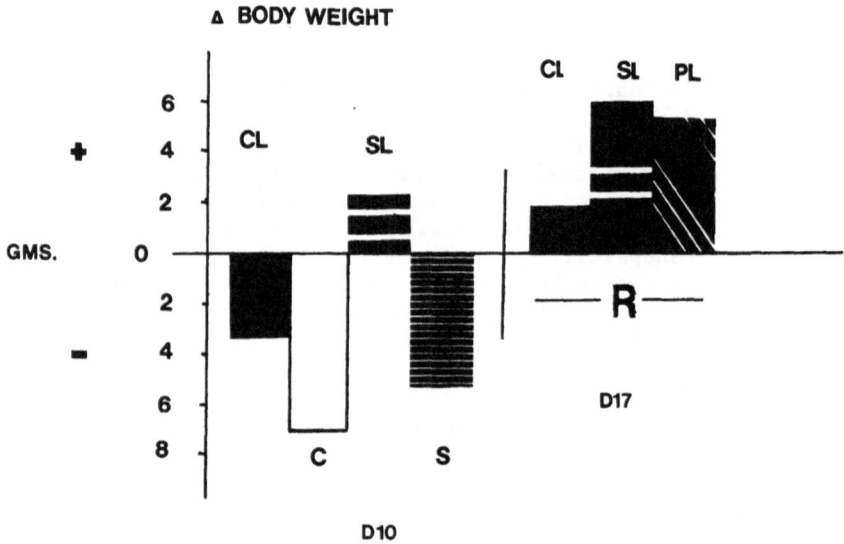

Fig. 1. Mean body weight change as a function of dietary fat (C, coconut; S, safflower; P, peanut oil) and lecithin (L) addenda. R indicates re-supplementation period.

During these two test periods, the intestinal weight of the total small bowel varied also with the type of fat fed (see Table 2). In the case of the coconut oil group, deprivation of the lecithin resulted in marked enlargement of the bowel, presumably an indication of dietary fat being accumulated and not transferred into the body by the mucosal cells. The wet weight of the intestines of animals given the lecithin addendum and coconut oil did not gain weight, indicating that they did not accumulate lipid.

In contrast, the safflower fed groups showed no detectable fat accumulation and transport problem. The weight of the small bowel was not noticeably heavier in the absence of dietary lecithin than in the reference group. The growth data tended to corroborate this, with the data for animals given safflower with lecithin (SL) gaining much more weight than the lecithin-deprived group (S), Table 2.

Liver weight generally varied inversely with the lecithin addendum in the diet, except in the case of the coconut oil diet group deprived of the lecithin supplement, Table 2. In this group the ability of the gut to absorb dietary lipid normally may have reduced the animal's hepatic fat store, hence weight, which decreased from 1.6 to 0.8 g. Safflower fed groups again afforded a sharp contrast with the data of the coconut groups. No inordinate amount of fat appeared to accumulate in the liver of the animals during these trials, if animals were fed safflower with phospholipid. In either case, the average liver weight decreased from 1.7 to 1.4 g. After being re-supplemented with lecithin, the liver weight increased to 1.3 g in the coconut group, approaching the 1.6 g seen in controls. Likewise, in the safflower group deprived of lecithin and then re-supplemented, the liver weight increased to 1.8 g versus 1.7 g for the safflower controls.

The peanut oil group (formerly fed safflower oil without lecithin) re-supplemented with dietary lecithin had an average liver weight of 1.8 g, similar to the weight observed in the safflower group re-supplemented with lecithin.

In these trials neither the type of dietary fat, nor the presence of lecithin markedly affected the volume of bile produced. Total bile flow and volume may respond to dietary features other than acyl composition and phospholipid level, Table 2. Animals consuming the safflower ration tended to produce somewhat larger bile volumes, though not significantly larger.

The mass of total bile salt sampled in the bile did not differ significantly in the coconut oil groups, Table 2. Though the total mass of bile salts in the gallbladder of lecithin supplemented or deprived groups did not differ, the deprived animals subsequently given the lecithin addendum for seven (7) days produced an average 50% more bile conjugate. The presence of lecithin in the safflower oil diets was not associated with increased bile output. A further addendum for one week did not result in additional bile salt output on average. Animals deprived of lecithin in the safflower group, and shifted to peanut oil during the seven-day lecithin re-supplementation period also produced about as much as the reference safflower oil group, and twice the mass produced by the coconut oil group.

Lecithin addendum to the diet influenced the bile salt profile of gallbladder bile. The presence of lecithin in the ration resulted in increased levels of glycine conjugates in the bile samples from animals fed either oil, Figures 2 & 3, and Table 3. Generally, the taurocholate level increased, or showed little change, if diet was fed without lecithin. After additional lecithin was introduced into the diet on day 11 for the seven-day period of re-supplementation (R), taurocholate levels decreased in both the coconut oil and safflower groups relative to the glycine conjugates, as indicated by the G/T ratio. This was also true in the group re-fed peanut oil. In the coconut group, the ratio decreased from 0.68 in the lecithin-fed group to 0.54 in the group deprived of lecithin; the level rose after lecithin was restored to the diet to 0.62. In the safflower oil group given lecithin the ratio decreased from 0.70 to 0.64; subsequent inclusion of lecithin in the diet resulted in a ratio of 0.90. Restoration of lecithin to a diet with peanut oil in previously deprived animals yielded a G/T ratio of 1.16. These data suggested that the safflower-fed animal was able to maintain function of the hepatic cell in producing normal bile in terms of the G/T better than the animal fed the saturated coconut oil.

Maintenance of normal levels of dihydroxy conjugates (GCDC, TCDC, & TDC) also depended upon incorporation of lecithin into the diet. Data also appear in Figures 2 & 3, and Table 3. In the coconut oil group, lecithin deprivation was accompanied by a 34% decrease in dihydroxy bile salts; in the safflower oil group a 26% decrement was measured. After the initial

Table 2. Body and Organ Weights

Diet:	COCO + LEC	COCO	COCO + LEC RESUPP	SAFF + LEC	SAFF	SAFF + LEC RESUPP	PEAN + LEC RESUPP
Tissue, grams							
Body	35.7±4.10	25.8±3.11	26.3±0.58	40.0±2.68	31.7±3.68	45.6±5.59	46.3±4.51
Gut [% BW]	4.0±0.9	5.6±1.2	4.8±0.8	4.2±0.3	4.4±0.7	3.9±0.2	3.9±0.3
Liver	1.6±0.2	0.8±0.2	1.3±0.4	1.7±0.1	1.4±0.2	1.8±0.3	1.8±0.2
Bile, ul	7.7±3.4	6.1±7.0	9.0±3.6	12±6.1	9.9±7.7	6.2±2.2	23±2.6
Bile salt. ug	63.6±22.3	62.3±38.7	93.0±30.7	111±40.9	126±17.2	121±66.0	128±58.0

See text for details of diets. All measurements at day 10, except RESUPP, which were at day 17. Values expressed as means ± standard deviations, with five to seven animals per group, except in the RESUPP groups that consisted of three to four animals.

ten-day deprivation period when lecithin was re-supplemented in the diets of coconut oil animals for seven days, the proportion of dihydroxy bile salts was not restored to normal values. In the coconut oil group fed lecithin, dihydroxy bile salts were 17.44% of the total, decreasing to 11.54% in the lecithin deprived group, and 9.64% in the group re-supplemented with lecithin. Safflower oil feeding resulted in lower proportions of dihydroxy salts under all conditions of lecithin feeding. The percentage of dihydroxy salts decreased from 9.07% in the group given lecithin to 6.66 in the deprived group, which rose to 7.12% in the re-supplemented group. Lecithin re-supplementation in the peanut oil group was accompanied by the highest observed value in these studies, with 8.60% of the bile salts present as dihydroxy bile conjugates. So, in addition to the influence of dietary lecithin upon the composition of bile with respect to the bile salt profile, the acyl composition of the dietary lipid also altered the relative proportion of various bile salts.

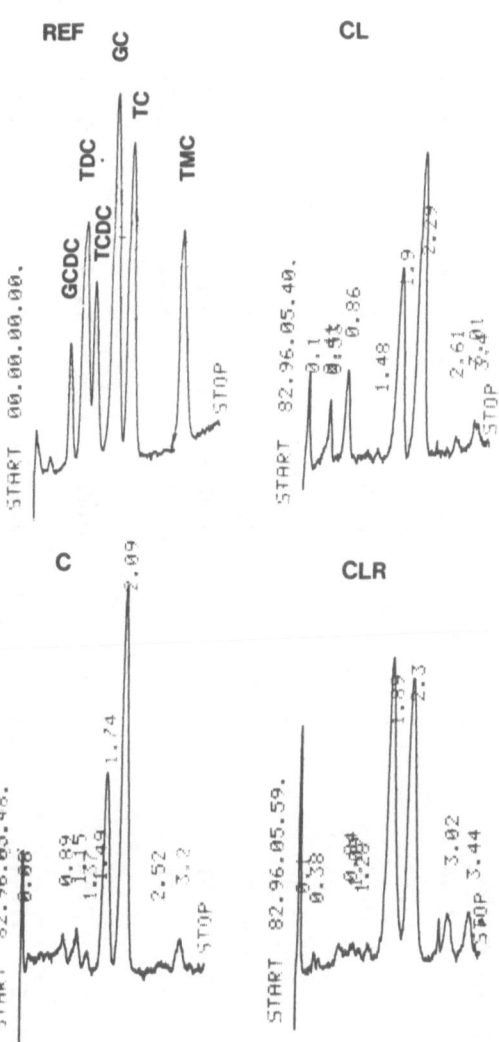

Fig. 2. Chromatograms of gall bladder bile of animals fed coconut oil with lecithin (CL), coconut oil without lecithin (C), or re-supplemented with lecithin (CLR). REF indicates a chromatogram of pure reference compounds.

Table 3. Bile Conjugate Composition

Diet:	COCO + LEC	COCO	COCO + LEC RESUPP	SAFF + LEC	SAFF	SAFF + LEC RESUPP	PEAN + L RESUPP
Conjugate°, weight %							
X	1.94	1.31	1.12	1.14	0.74	0.52	0.52
GCDC	14.6	5.61	2.47	5.03	1.72	2.10	5.47
TDC	0.80	2.58	2.55	2.58	2.93	2.21	1.02
TCDC	2.04	3.35	4.62	1.46	2.01	2.81	2.11
GC	25.9	29.4	35.8	36.3	37.5	45.2	48.1
TC	49.9	54.0	47.6	42.3	41.3	37.8	16.3
Y	2.51	1.80	3.92	5.67	7.06	5.56	20.6
TMC	2.31	1.88	1.92	5.44	6.78	4.29	5.90
G/T	0.68	0.54	0.62	0.70	0.64	0.90	1.16
Di-OH, %	17.44	11.54	9.64	9.07	6.66	7.12	8.60

° X, Y: identities unknown; GCDC: glycochenodeoxycholate; TDC: taurodeoxycholate; TCDC: taurochenodeoxycholate; GC: glycocholate; TC: taurocholate; TMC: tauromuricholate; G/T: mass ratio of glyco- to tauro- conjugates; Di-OH: sum of dihydroxy conjugates expressed as per cent of total conjugates. Values of conjugates are means of five to seven animals, except RESUPP, which represent three to four.

The nature of dietary fat fed modulated the composition of bile conjugates in gerbil gall bladder bile. When unsaturated safflower oil (or peanut oil) was fed, the proportion of glycine conjugates increased. Based upon growth, organ weights, and general vigor of these animals, this glycine-enriched bile was always associated with better health. Since the actual chemical composition of the bile responded to alterations in the diet, and the bile is synthesized by the hepatic cell, the alterations in dietary lipid resulted in modified liver cell function, if either lecithin or fatty acid input were changed. Coconut oil feeding was associated with glycine conjugate-poor bile.

The liver cell apparently used taurine conjugates to try to maintain bile functionality in the absence of adequate lecithin in the diet. Though no measurable effect of dietary lecithin upon bile volume was observed in these trials, the presence of lecithin in the diet appeared to facilitate synthesis and use of glycine conjugates, possibly releasing taurine for other sulfur needs.

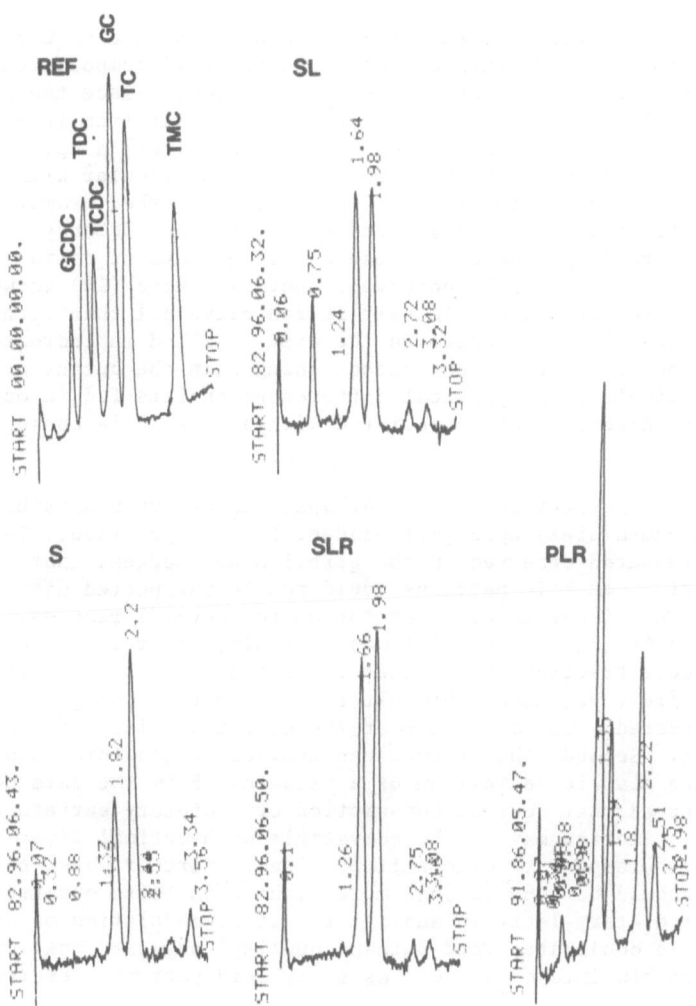

Fig. 3. Chromatograms of gall bladder bile of animals
         fed safflower oil with lecithin (SL), safflower
         oil without lecithin (S); or, re-supplemented
         with lecithin (SLR), or lecithin and peanut
         oil (PLR). REF indicates a chromatogram of
         pure reference compounds.

175

The animal substituted a taurine-rich bile salt profile when deprived of dietary lecithin. The pKa of taurine is much lower than the pKa of glycine, so a mixture enriched with taurine conjugates would be expected to have some enhancement of surface activity. These data are not presented in this report. That the dietary inclusion of lecithin may also have increased the phospholipid content of the bile has not been part of this paper.

Alteration of quality of the dietary lipid in these studies resulted in changes in the state of hydroxylation of the bile conjugates. These data suggest that dietary lecithin apparently mediates at least partial regulation of hydroxylation of the bile acids, whether directly or indirectly. The lecithin deprived liver cell shuttled more cholesterol nucleus into the tri-hydroxy conjugates of both GC and TC. It is presumed that this would enhance the hydrophilicity of the bile under suboptimal conditions.

Discussion

Bile salt composition was expected to change with the nature of dietary fat, including both the lecithin addendum and the acyl composition. The data presented in these trials supported this expectation. Since the observation of Burr & Burr (9) that omission of polyunsaturated fat from the diet resulted in skin lesions and deranged fat processing in the liver, a correlation between structure and function has been known, in particular with respect to transport of fat out of the liver. Further, the rather dramatic changes of serum cholesterol levels demonstrated by Hegsted and McGandy (6) in short-term human feeding studies suggested that effects of dietary lipid upon bile quality might also be observed. Bothem & Boyd (10) showed that rat hepatocytes produced less bile salt if fed olive oil than corn oil, or no fat supplement. A fat addendum in the diet resulted in increased cholate and chenodeoxycholate. We did not detect changes in the output of bile salt. The relatively large physical differences in fats fed in our studies at fixed levels, however, were associated with altered bile conjugate patterns.

A question of greater interest, perhaps, focuses upon possible effects of lecithin in human diets upon gall bladder bile composition. The changes that we have presented observed in the gerbil model suggest that such modifications of human bile patterns would not be unexpected with dietary manipulation. This issue deserves attention for several reasons. First, a mere three per cent (3%) of the diet was lecithin, about 14% of dietary fat, which was a modest fraction of the diet in contrast to the fat effect which was 20% of the diet, yet measurable alterations in the conjugated bile salt composition occurred. The amplitude of the effect was large for a minor fat constituent. Second, the tissues can synthesize lecithin, and hence it does not fit the classic definition of a vitamin. From the data here presented, we inferred that routine consumption of a dietary pattern typical of that eaten in America may not be compatible with optimal liver function, particularly with respect to bile salt profiles. Further work, both chemical and physical, should be done to shed more light upon these observations. That changes of dietary lecithin and fat result in modulation of quality of gall bladder bile conjugates (and perhaps quality) suggests that dietary patterns of lecithin intake, as well as fatty acid patterns deserve closer examination.

As pointed out above, a large number of adults suffer from cholestasis and gallstones, with an additional 800,000 cases reported per year (13). Also, special groups such as the cystic fibrotic, plagued with a genetic defect and abnormal bile composition, has severely imparied fat digestive capacity. This results in impaired growth and other complications. As in the data reported by Harries, et al. (5), the bile conjugates of the cystic fibrotic patient were reduced in amount, with an altered glycine conjugate content, as well as reduced dihydroxy conjugate levels. These subjects have been helped with pancreatic enzyme supplements. Neither their tissue fatty acid profiles, nor their bile slat composition has yet been restored to normal. Could routine dietary phospholipid supplementation offer a means to bring their bile quality, digestion, and growth pattern toward normal? If so, their quality of life would benefit, especially in terms of resistance to disease: as their weight decreases they lose resistance to infection which often becomes fatal. In the case of the adult suffering with cholestatic bile, the condition at best may cause pain after a meal and at worst surgery to remove stones. Currently, a non-invasive sonic treatment may afford some relief. The ideal strategy would be the strategy that prevents stone formation.

A third reason to consider dietary patterns and the lecithin effect upon bile quality and general liver and small bowel vigor seen in these data when lecithin was included in the diet, is that in spite of government recommendations for the past ten years, namely the Dietary Goals and later Guidelines, the typical American continues to eat a fat-rich, surfactant-poor diet. One wonders whether along with appropriate fat, a lecithin addendum into the diet would also be accompanied by improved biliary quality, digestion, and general health. One consequence of this might be a reduced national health bill. Such speculation must give way to evidence from further feeding studies with lecithin, the vitamin-like natural emulsifier and membrane element that has been largely removed from the food supply. Current statistics about cholestasis and associated disorders suggest that the fat intake habits now practiced may condition a dietary need for lecithin supplementation for optimal health. However, before enlightened judgments could be made about appropriate lecithin levels in the current diet, actual data about lecithin and fat intake patterns should be measured. Older estimates do not provide a trustworthy basis for such judgments, since dietary patterns change and newer, more precise and reliable analytical methods exist for measuring dietary lecithin intake. These data point to the need for more dietary studies, judgment and action.

References

1. D. M. Hegsted, Nutrition: The Changing Scene, Nutr. Rev. 43:357 (1985).
2. M. C. Bateson, Dissolving Gallstones, Br. Med. J. 284:1 (1982).
3. H. G. Parsons, et al., Energy Needs & Growth in Children with Cystic Fibrosis, J. Ped. Gastroent. & Nutr. 2:44 (1983).
4. P. Farrell, et al., Occurrence and Effects of Vitamin E Deficiency in Cystic Fibrosis, J. Clin. Invest. 60:233 (1977).
5. J. F. Harries, et al., Intestinal Bile Salts in Cystic Fibrosis, J. Dis. Child. 54:19 (1979).
6. D. M. Hegsted & R. Mc Gandy, Qualitative Effects of Dietary Fat & Cholesterol upon Serum Lipids in Man, in: The Role of Fats in Human Nutrition, A. Vergroesen, ed., Academic Press, New York (1975).
7. E. K. Wong, et al., Influence of Dietary Lecithin on Hyperlipidemia in Rhesus Monkeys, Lipids 15:428 (1980).
8. S. Levin & J. Touchstone, Reverse Phase TLC for Bile Acids in Rat Bile, in: Advances in Thin Layer Chromatography, J. Wiley, New York (1982).
9. G. O. Burr & M. M. Burr, Essentiality of Lipid in the Diet, J. Biol. Chem. 86:587 (1930).
10. K. Bothem & G. S. Boyd, The Effect of Dietary Fat upon Bile Salt Synthesis in Rat Liver, Biochim. Biophys. Acta 752:307 (1983).
11. J. D. Lloyd-Still, et al., Essential Fatty Acid State in Cystic Fibrosis & the Effects of Safflower Oil Supplementation, Am. J. Clin. Nutr. 34:7 (1981).

12. C. C. Roy, et al., Abnormal Biliary Lipid Composition in Cystic Fibrosis, New Engl. J. Med. 29:1301 (1977).
13. K. Taylor & L. Anthony, Nutritional Aspects of Liver & Biliary Disease, Clinical Nutrition, McGraw Hill, New York (1983).

SUMMARY OF PANEL DISCUSSIONS

The 4th International Colloquium on Lecithin in Chicago ended
with a stimulating and provocative series of four simultaneous panel
discussions, which involved all the participants and attendees in the
Colloquium. Each discussion group was led by a moderator, and
explored a list of questions generated by the group, based on issues
and concerns around a specific topic. Subsequently, all the
attendees reconvened, and each of the moderators provided a brief
summary of the questions which were covered in their respective panel
discussions.

The following are general capsules of the four panel
discussions. The summary by Dr. Ansell was prepared posthumously,
and it is a modified transcript from a tape recording of his
presentation.

PANEL DISCUSSION #1: NOMENCLATURE, QUALITY CONTROL AND
STANDARDIZATION OF COMMERCIAL LECITHIN

**Moderated and Summarized by: Michael Schneider, Lucas Meyer GmbH &
Co., Hamburg, Germany.**

1. Nomenclature

There is no need for a clarification of the nomenclature of
single phospholipids, since definitive recommendations by the IUPAC-
IUB Commision on Biochemical Nomenclature were made in 1976. The
discussion focused, therefore, on the definition of the term
lecithin. In the scientific literature the trivial expression
lecithin is still used as a synonym for phosphatidylcholine
(sn-1,2,-diacylglycero-3-phosphocholine) whereas in the
pharmacopeial, legislative and commercial senses the term **lecithin** is
used for the natural mixtures of different phospholipids from
different origins. Moreover, even within these terms there still is
some ambiguity. Therefore, the participants agreed on the necessity
for a universally accepted definition, so as to avoid severe
misunderstandings between consumer and producer.

2. Quality Control

Just as there is a desire for uniform nomenclature, there was
considerable emphasis on the need to devise uniform methods of
analysis, be it with respect to phospholipid determination, or with
respect to safety-related problems (toxicological, microbiological,
problem of impurities, etc.).

3. Standardization

The definition of the term **purity** was discussed in the context of standardization. The panel felt strongly that the aim of standardization should be: a) the uniform determination and description of products; and b) the use of the same terminology by all investigators.

Conclusion

In order to create the widest possible platform for all interested parties, it was decided to form an International Study Group on Lecithin which would deal with the above mentioned issues, as well as related topics. The composition of this Study Group should reflect all interests. Dr. Schneider offered to contact all interested persons, and to arrange for a meeting of such a Study Group, early in 1987. The outcome of this study group will be presented at the next International Colloquium on Lecithin, which is scheduled for sometime in 1988.

PANEL DISCUSSION #2: THERAPEUTIC AND NUTRITIONAL RELEVANCE OF PHOSPHOLIPID ADMINISTRATION

Moderated and Summarized by: Israel Hanin, Loyola University of Chicago Stritch School of Medicine, Maywood, Illinois, USA.

Four questions were posed for discussion by this panel:

Question #1: "Is there a recommended upper dose of lecithin that should be used, and are there side effects following lecithin ingestion that one should be concerned about?"

The group agreed that terminology must first be defined in this case. When one speaks of **lecithin**, is one referring to the pure substance, the granulated form, or to lecithin emulsions? If it is the "pure" substance, what is the extent of its purity? Once these differences are defined, one could begin to address the issue of upper limits of lecithin to be used.

Studies from rats appear to indicate that this specie is able to tolerate 4 grams lecithin /kg body weight, with no apparent side effects. Because rats have a higher metabolic rate than humans, it is difficult to translate the data to human terms. Side effects which have been noted in humans include gastrointestinal disturbances, diarrhea, weight gain, and some incidences of rash and headaches.

It was suggested, based on available clinical information, that one should not exceed a dose of 10 grams of lecithin/day in a human subject. This refers to the total amount of pure lecithin to be used. Thus, if for example, one were to use a granular preparation in which only 20% of the content consisted of lecithin, then 50 grams of the preparation would have to be ingested. In this case, some caution should be taken in identifying the contents of the remaining 80% of the preparation, and in assuring that intake of these ingredients in large quantities is not posing unnecessary complications for the user.

Question #2: "Should lecithin be used prophylactically on a regular basis?"

With our present available information there is no way of knowing whether long-term administration of lecithin at low doses has either a good, or detrimental effect. It was suggested that prophylactic use of lecithin should not be necessary if people were to eat a proper diet. However, we do not know what such a "proper diet" would be. A well-balanced lecithin-rich diet has not yet been formulated. Hence, as in the case of the previous question, no clear-cut suggestions were made by the panel group in attempting to address the issues inherent in this question.

Question #3: "What is the efficacy of lecithin in the treatment of memory disorders?"

The reports by Dr. Sorgatz on the one hand, and by Dr. Growdon on the other, clearly illustrated the current controversy in the field regarding the use of lecithin in the treatment of memory disorders. While Dr. Growdon's presentation demonstrated, in the course of a review of the literature, a general lack of effect of lecithin, Dr. Sorgatz's study showed statistically significant evidence for improvement in memory when certain subjects consumed lecithin. Certain questions were raised in the discussion of these results, in an attempt to resolve this difference.

For instance, the subjects in Dr. Sorgatz's study were from a low income segment of the population. Could their diet have had some effect on the observed results? Also, the age range of this population of volunteers was between 45-55 years old. Perhaps there is an age-related phenomenon in the role of lecithin and memory? Finally, the formulation of the material presented to the population studied was in an ethanol/water suspension. Could this combination have had some special additive effect in the treatment response of Dr. Sorgatz's volunteers? Could the alcohol conceivably have caused a change in absorption efficiency of the substance in the body?

Question #4: "Active Lipid (AL 721)"

Discussion focussed on various aspects of AL 721, the material described by Dr. Shinitzky earlier in this Symposium (see chapter by Shinitzky and Haimovitz).

This substance (consisting of a suspension of 70% neutral lipid, 20% phosphatidylcholine and 10% phosphatidylethanolamine) is in micellar form when it is administered. One question was raised as to the form the lecithin might be in when it reaches the target organ where it exerts its action. While it is difficult to determine the answer to that question, Dr. Shinitzky stressed the fact that the radiolabelled compound was distributed to different parts of the body (lymph. blood, brain), and that an expected immune response was reproducibly observed in the subjects tested. Hence, irrespective of the form of the compound, it apparently reaches its target in a specific and selective manner.

It is interesting to note that all of the patients who responded to AL 721 treatment were over 75 years of age. Younger subjects did not respond.

In response to a question regarding the difference between AL 721 and liposomes, it was stated that liposomes are much more stable than AL 721. On the other hand, the latter is more biologically active of the two.

PANEL DISCUSSION #3:   NEW DRUG PREPARATIONS INVOLVING PHOSPHOLIPIDS

Moderated and Summarized by: G. Pepeu, University of Florence, Italy.

The following questions, prepared by the moderator of the panel, or suggested by the participants, were brought up by consensus, for general discussion:

1. Phospholipids as drug carriers or vehicles

   a. What are the advantages of liposomes over emulsions?
   b. What drugs are improved by inclusion into liposomes?
   c. What are the best phospholipids for preparing liposomes?
   d. What, if any, are the liposome side-effects?

2. Phospholipids as drugs

   a. What are the indications for use of phospholipids as drugs?
   b. Which preparations of phospholipids could be used as potential therapeutic substances?

An impromptu discussion on the first point revealed a large interest in the possibility offered by liposomes as drug carriers. It was mentioned that in the U.S. only 80 patents on the use and preparation of liposomes have been granted. It also was pointed out that little information is presently available on liposome biological activities, duration and fate. With regard to the biochemical makeup of liposomes, according to the discussants, mixtures of cholesterol and phosphatidylcholine are the most common components of liposomes.

Several interesting observations were reported in the discussion. Dr. Fareed mentioned that inclusion into liposomes alters the pharmacological properties of heparin. It is not known whether the action of other drugs, such as antitumor agents, are also modified by liposome inclusion. Liposomes penetrate into the cells and, according to investigators of the Johnson Wax Company, they remain much longer in skin fibroblasts than in microphages, by which they are demolished.

The general consensus was that liposomes are not inert carriers. Their direct activity depends on their size, route of administration, and their pharmacological and toxicological effects need to be carefully investigated. Conversely, no agreement was reached as to which drugs are really improved by inclusion into liposomes. The hope that liposome formulation might increase drug selectivity and reduce toxicity still needs to be confirmed by controlled studies.

The use of phospholipid emulsions as drug vehicles appears to be much simpler, when compared to the liposomes. As reported elsewhere in this Colloquium by Dutot, lecithin emulsions are well tolerated by patients. Emulsions protect the tissues from local irritant properties of some drugs. On the other hand, the inertness of emulsions depends on their lipid composition. As presented by Toffano at this Colloquium, emulsions of phosphatidylserine exert well defined and useful pharmacological effects.

The discussion then moved on to the fate of the administered phospholipids and their fusion with biological membranes. The second point presented for discussion, phospholipids as drugs, was therefore taken up.

So far the phospholipids which have been used as drugs are: phosphatidylcholine (lecithin), phosphatidylserine, dipalmitoyl-phosphatidylcholine, the "active lipid" proposed by Shinitzky and Haimovitz, and the phospholipid cytidinediphosphocholine.

The latter is used as a drug in Japan and several European countries, for the treatment of a variety of medical conditions ranging from stroke, to the aging brain. Some of the pharmacological and clinical evidence is interesting but it has not been clearly established whether at therapeutic doses this compound really acts as a phospholipid precursor. Dipalmitoylphosphatidylcholine is used as a lung surfactant. For the three remaining phospholipids, the strict relationship between the drug and its pharmaceutical form emerged again. Finally, too little was known about the composition of "active lipid" to allow a knowledgeable discussion (but see chapter by Shinitzky and Haimovitz in this book).

Dr. Growdon, referring to some of the data presented by Toffano and coinvestigators for phosphatidylserine, and to some, as yet unpublished data by Mervis and his coworkers, with regard to choline, called attention to the apparent ability of some phospholipids to prevent aging, at least in rodents. Phosphatidylserine, for instance, prevents the morphological changes induced by age in hippocampal neurons. At present it is not known whether this effect is widespread, or whether it involves only a limited number of cells. With regard to a proposed mechanism of action of phosphatidylserine in the prevention of aging, Dr. Toffano suggested, as a working hypothesis, two mechanisms of action: a) specific phospholipids may fuse with membranes, restoring their properties altered by age; and b) phospholipids may facilitate the synthesis, release, and action of trophic factors (e.g. nerve growth factor). Much investigation will be necessary to define more precisely the effect of phospholipids on aging, and to understand its mechanism. Nevertheless, this is one of the most promising fields of phospholipid research.

The panel also brought up the question of which substances should be used as controls in pharmacological research on phospholipids. Must the effects of a given phospholipid be evaluated against those of its individual components or of other, possibly inert, phospholipids? No definite answer was indicated, but it was suggested that one should carefully select the control substance case by case, without forgetting that the pharmacological actions of phospholipids may depend strongly on the physico-chemical properties of their pharmaceutical preparations.

Moderated and Reported by: G. Brian Ansell, The Medical School, University of Birmingham, U.K. (adapted from a tape-recorded transcript).

During the panel discussion we first started by evaluating our present understanding of the various phospholipid pathways. We wondered whether any new pathways for phospholipid metabolism had emerged briefly. Apart from the generally accepted pathway for the synthesis of phosphatidylserine another pathway has been suggested. This one uses phosphatidic acid and nickel as a cofactor. This pathway erupted about five years ago and subsequently disappeared without a trace. It would be a very unusual reaction, and one wonders that if it did exist, whether there is a possibility of similar reactions affecting other phospholipids.

In further discussion of phosphatidylserine, we decided that the sole fate of this substance must be in the synthesis of phosphatidyl-ethanolamine. It still is one of the more mysterious phospholipids.

We then discussed glycerophosphocholine and whether, in fact, there is a synthetic pathway for it, whether it might just be a simple breakdown product in vivo, and what this might mean for the general metabolism of choline lipids in the brain in general.

This led to a discussion of the new technique which is emerging for studying phospholipid metabolism in vivo: nuclear magnetic resonance. Such a technique is obviously essential, if you are going to look at studies of metabolism in humans in health and disease. Nevertheless, it is fair to say that we are a little skeptical about this approach. It is something that needs careful study. At the moment, it can only be used in observing water-soluble phosphorus compounds; it cannot be used in observing phospholipids.

The approach of using nuclear magnetic resonance, if and when it is perfected for observing phospholipids, might be useful in studying the ever-perennial problem of turnover and what it means. Turnover is not as fashionable in vivo as it once was; but it is obviously very important. As one member of the panel pointed out, a given phospholipid may have numerous different turnovers, depending upon its molecular species and location. It is very difficult to sort this out because it may be that there are some small pools of particular phospholipids which have a particular function.

This led to discussion of the function of phospholipids and whether there is any sort of a second messenger possibility. We all know about the so-called PI response. We skated briefly over methylation. Dr. Horrocks expanded on a theory that there may be an involvement of choline plasmalogens in new second messenger function. He also pointed out that there was a very interesting trans-arachidonillation from an energy-independent transfer of arachidonate to choline plasmalogen, which could then be rapidly released. The release of arachidonic acid is a very important reaction in tissues. When this can be activated by certain agonists, then we have the possibility of a second messenger system.

The next topic discussed was the contribution that molecular biology will make to phospholipid research. It is clear from other fields, particularly the receptor field, that the contribution of molecular biology in the cloning and sequencing of important components is quite astounding. Dr. Vance raised the hope that certain enzymes in phospholipid metabolism may soon be purified and cloned using a DNA library, in order to obtain these enzymes in larger amounts. This would make it possible, for example, in a particular enzyme, to find the phosphorylation sites. There are, however, problems with the purification of enzymes involved in phospholipid metabolism. Many of them are just intracellular phospholipases, and we do not know much about their molecular weight. There thus is a considerable amount of work to be done, at an extremely sophisticated level.

Finally, we turned our attention to the problem of absorption of micelles. The general consensus was that only a very small proportion of phospholipids, or lipids at all, can be absorbed intact. It was also generally felt that there is not much information presently available in favor of micelle absorption per se. Perhaps a new, modified micelle will soon be developed, which would be well absorbed in vivo. At this stage in time, however, such a substance is not available.

EDITORIAL NOTE: In the context of each of these four panels, all attempts have been made to be as accurate as possible in reconstructing the views, reports and their conclusions. It should also be noted that all these reports reflect the point of view of the specific participants in the individual discussion groups, and thus do not necessarily represent the view of all the participants in the Colloquium, let alone the field of lecithin investigators on the whole. This chapter does, nevertheless, provide an indication of the kinds of questions which were raised, and thus gives the reader some sense of the issues of interest at the time of the Colloquium.

PHOTOGRAPH OF SPEAKERS IN THE COLLOQUIUM

(Left to Right)

FRONT ROW: M. Shinitzky, K.-H. Schmidt, M. Schneider,
             T.R. Watkins, G. Pepeu, D.E. Vance,
             B. Akesson, J.K. Blusztajn, G. Toffano,
             S.H. Zeisel, W. Feldheim.

BACK ROW: G.B. Ansell, I. Hanin, F.J. Martin,
            G. Dutot, F. Paltauf, L.A. Horrocks,
            C. Ratledge, J. Hayward

Missing from the photograph: J.H. Growdon, H. Sorgatz

LIST OF PARTICIPANTS

ÅKESSON, Björn   University Hospital
                                 Lund/Sweden

ANDRES, Achim   Nattermann Chemie GmbH
                                 Cologne/F.R. Germany

ANSELL, G. Brian   University of Birmingham Medical
                                 School, Birmingham/G. Britain

BARNES, Diane   Loyola University of Chicago
                                 Maywood, IL/U.S.A

BARNES, Stanley   Lucas Meyer Inc.
                                 Decatur, IL/U.S.A.

BEERY, Kenneth E.   Central Soya Company, Inc.
                                 Fort Wayne, IN/U.S.A.

BERRY, Ira R.   Pharmacaps, Inc.
                                 Elizabeth, NJ/U.S.A.

BLAIR, Sidney J.   Loyola University of Chicago
                                 Maywood, IL/U.S.A.

BLESSA, Roberto   Sanbra, Soc. Algodoeira de Nordeste
                                 Brasileiro S.A., Sao Paulo/Brazil

BLUSZTAJN, Jan K.   Massachusetts Institute of Technology
                                 Cambridge, MA/U.S.A.

BRANDT, Paul W.   American Feeds & Livestock Co., Inc.
                                 Long Grove, IL/U.S.A.

BURSENS, William   UCB S.A.,
                                 Brussels/Belgium

CAPOZZE, Jack C.   University of Illinois Medical School
                                 Chicago, IL/U.S.A.

CHARAF, Ursula K.   S.C. Johnson & Son., Inc.
                                 Racine, WI/U.S.A.

COTTER, Richard   Travenol Laboratories
                                 Round Lake, IL/U.S.A.

COYNE, Erwin   Loyola University of Chicago
                                 Maywood, IL/U.S.A.

CURRY, Jim	Riceland Foods, Inc. Little Rock, AR/U.S.A.
DAVIS, Edward T.	Nutrition Headquarters, Inc. Carbondale, IL/U.S.A.
DAVIS, Joseph R.	Loyola University of Chicago Maywood, IL/U.S.A.
LA DROITTE, Philippe	Dietetique & Sante, S.A. Revel/France
DUENAS, Pedro	Lucas Meyer S.A. Barcelona/Spain
DUTOT, Guy	Laboratoires Cernep Synthelabo Le Plessis Robinson/France
FANNING, Frank G.	The Fanning Corp. Chicago, IL/U.S.A.
FAREED, Jawed	Loyola University of Chicago Maywood, IL/U.S.A.
FELDHEIM, Walter	Institute of Human Nutrition University of Kiel, Kiel/F.R. Germany
FOX, Johannes M.	A. Nattermann & Cie., GmbH Koln/F.R. Germany
FRANCOIS, Roger	Lucas Meyer France S.A. St. Maur des Fossés/France
FRANKFATER, Allen	Loyola University of Chicago Maywood, IL/U.S.A.
FRANZMAIR, Rudolf	Chemie Linz AG Linz/Austria
FRIEDMAN, Alexander H.	Loyola University of Chicago Maywood, IL/U.S.A.
FUKUSHIMA, Minoru	The Institute for Early Development of Japan, Inc. Tokyo/Japan
FUNG, Leslie	Loyola University of Chicago Maywood, IL/U.S.A.
GODIN, Dominique	UCB S.A. Brussels/Belgium
GOETZ, Volker	Roland Arzneimittel GmbH. Hamburg/F.R. Germany
GRAF, Tilman	Lucas Meyer GmbH & Co. Hamburg/F.R. Germany

GRAHAM, Harold N.	Thomas J. Lipton, Inc. Englewood Cliffs, N.J./U.S.A.
GUENTHER, Bernd R.	A. Nattermann & Cie., GmbH Köln/F.R. Germany
HANIN, Israel	Loyola University of Chicago Maywood, IL/U.S.A.
HASEGAWA, Mineo	Q.P. Corporation Tokyo/Japan
HAYNES, Lynn C.	The Upjohn Co. Kalamazoo, MI/U.S.A.
HAYWARD, James	State University of New York Stony Brook, N.Y./U.S.A.
HORROCKS, Lloyd A.	The Ohio State University Columbus, OH/U.S.A.
HOWARD, Norman B.	Oxford, OH/U.S.A.
HEISER, Jens	Lucas Meyer GmbH & Co. Hamburg/F.R. Germany
KARTINOS, Nicholas T.	Austin Chemical Chicago, IL/U.S.A.
KATO, Noboru	Hohnen Oil Corp. Ltd. Tokyo/Japan
DE KERCHOVE DE DENTERGUEM, Henri	Sanbra - Soc. Algodoeira do Nordeste Brasileiro S.A. Sao Paulo/Brazil
KINDEL, Gisela	Loyola University of Chicago Maywood, IL/U.S.A.
KOMODA, Namoru	Hohnen Oil Co. Ltd. Tokyo/Japan
KOPELMAN, Paul	Zivi, Broitman & Kopelman Chicago, IL/U.S.A.
LEBOURD, Pierre	Lucas Meyer France S.A. St.-Maur-des-Fosses/France
LEVENTER, Steven	Loyola University of Chicago Maywood, IL/U.S.A.
MACDONALD, Kevin	American Lecithin Company Atlanta, GA/U.S.A.
MANTEUFFEL, Mary Druse	Loyola University of Chicago Maywood, IL/U.S.A.

MARSH, R.A.	RHM Foods, London/G. Britain
MARTIN, Francis J.	Liposome Technology, Inc. Menlo Park, CA/U.S.A.
McCARTHY, James P.	Central Soya Company, Inc. Fort Wayne, IN/U.S.A.
McGUIRE, Patricia	Loyola University of Chicago Maywood, IL/U.S.A.
MEYER, Lucas	Lucas Meyer GmbH & Co. Hamburg/F.R. Germany
MEYER, Jeremy	Mars UK Ltd. Slough/G. Britain
MILLER, Gregory	Kraft, Inc. Glenview, IL/U.S.A.
MORRISON, Lester M.	Institute for Arteriosclerosis Research Los Angeles, CA/U.S.A.
MOWLES, Don	Abbott Laboratories Abbott Park, IL/U.S.A.
NIEUWENHUYZEN, Willem van	Croklaan B.V.-Unimills Wormerweer/The Netherlands
OKADA, Masuo	Sanbra - Soc. Algodoeira do Nordeste Brasileiro S.A., Sao Paulo/Brazil
OSHIDA, Kazuo	Q.P. Corporation Tokyo/Japan
PALTAUF, Fritz	Technical University of Graz Graz/Austria
PARTH, Karl	Karl Parth Vienna/Austria
PATTERSON, Richardson J.	Microfluidics Corp. Newton, MA/U.S.A.
PAVLINA, Tom	Travenol Laboratories Round Lake, IL/U.S.A.
PEPEU, Giancarlo	University of Florence Florence, Italy
PFEIFFER, Ulrike	Lucas Meyer GmbH & Co. Hamburg/F.R. Germany

PINKPANK, Axel	Lucas Meyer, Inc. Decatur, IL/U.S.A.
POTTER, Pamela E.	Loyola University of Chicago Maywood, IL/U.S.A.
PRESTON, Mike A.	Lucas Meyer (UK) Ltd. Chester/G. Britain
RAMABADRAN, K.	Erie Casein Company, Inc. Rochelle, IL/U.S.A.
RATLEDGE, Colin	University of Hull Hull/G. Britain
V.D REEP, Franziskus M.	Natufood BV, Harderwijk/The Netherlands
ROHDE, Peter	Lucas Meyer GmbH & Co. Hamburg/F.R. Germany
ROSA, James	New Foods, Villa Park, IL/U.S.A.
ROSENBERGER, Henry K.	Nutrition Headquarters, Inc. Carbondale, IL/U.S.A.
DE SAINT-BLANQUAT, Guy	Inserm Toulouse/France
SANDERSON, James	Kraft Inc., Glenview, IL/U.S.A.
SCHMIDT, Karlheinz	University of Tuebingen Tuebingen/F.R. Germany
SCHNEIDER, Michael	Lucas Meyer GmbH & Co. Hamburg/F.R. Germany
SHINITZKY, Meir	The Weizmann Institute of Sciences Rehovot/Israel
SILVERMAN, Craig	Loyola University of Chicago Maywood, IL/U.S.A.
SZUHAJ, Bernard F.	Central Soya Company, Inc. Fort Wayne, IN/U.S.A.
SKULTE, Axel B.	Lucas Meyer srl Vigonza, Pd/Italy
SORGATZ, Hardo	Technical University of Darmstadt Darmstadt/F.R.G.

STAHOPOULOS, Vicki	Northwestern University Evanston, IL/U.S.A.
STECHER, Walter	A. Nattermann & Cie. GmbH Köln/F.R. Germany
STERN, Paula H.	Northwestern University Evanston, IL/U.S.A.
TOFFANO, Gino	Fidia Research Laboratories Abano Terme, Pd/Italy
VANCE, Dennis E.	AHFMR Faculty of Medicine University of Alberta, Edmonton, Alberta/Canada
WATKINS, Tom R.	City University of New York New York, N.Y./U.S.A.
WHEELER, Edward L.	RJR Nabisco, Inc. West Hanover, N.J./U.S.A.
WEHRMACHER, William H.	Loyola University of Chicago Maywood, IL/U.S.A.
WOORALL, Charles T.	Central Soya Company, Inc. Fort Wayne, IN/U.S.A.
WULFERT, Ernst	UCB S.A. Brussels/Belgium
YAGI, Takashi	Showa Sangyo Co., Ltd. Tokyo/Japan
YOSHII, Yuichi	D.V. Preston, Inc. Yokohama/Japan
V.D. ZANDE, Andre	Lucas Meyer B.V. The Hague/The Netherlands
ZEISEL, Steven H.	Boston University School of Medicine Boston, MA/U.S.A.
ZHANG, Yin	Loyola University of Chicago Maywood, IL/U.S.A.
ZIEGELITZ, Rüdiger	Lucas Meyer GmbH & Co. Hamburg/F.R. Germany
ZIVI, Walter	Zivi, Broitman & Kopelman Chicago, IL/U.S.A.

INDEX

Rat (continued)
  brain lipids, 6
  and lecithin, 180
  and phosphatidylserine, 139–141
Rhizopus arrhizus, 8
Rhizopus delemar, 38
Rhodotorula, 28

Saccharomyces, 28–30
S-adenosylemethionine, 98
Safflower oil, 168–175
Schizophrenia, 126
Sclerosis, amyotrophic, lateral, 125
Serine, 140
Short bowel syndrome, 67–70
Soybean oil, 66–67, 101–106
Sphinogomyelin, 49, 108, 114
Sphingophospholipid, defined, 1
Spirulina maxima, 19–20
Spirulina platensis, 20
Striatum of rat brain, 89,91
Survival, cellular and pshopholipids,
      85–94

Tardive dyskinesia, 122–123
  and choline choline, 123
Targeting, 55
Taurine, 175–176
Thrombin, 5
Tourette syndrome, 125
Transduction of signal, 3, 6
Translocation hypothesis, 78–79
Transmethylation, 98
Triacylglycerol, 62, 63, 66, 67
Trichosporon, 28, 29
Triglyceride, 102, 104, 155, 157,
      158
Triolein, 69–70
  breath test, 69–70

Vesicle, 49–52 *see also* Membrane

Yeast, 28–31